I Am Taurus

I Am Taurus

Stephen Palmer

IFF
BOOKS

Winchester, UK
Washington, USA

The Girl with Two Souls
978-1710466249 (2016)

The Girl with One Friend
978-1670692122 (2016)

The Girl with No Soul
978-1676182245 (2016)

Tommy Catkins
978-0995752269 (2018)

The Autist
978-1795806749 (2019)

The Conscientious Objector
978-1673919356 (2019)

Tales from the Spired Inn
978-1912950423 (2019)

Woodland Revolution
979-8627105833 (2020)

JOHN HUNT PUBLISHING

First published by iff Books, 2024
iff Books is an imprint of John Hunt Publishing Ltd., No. 3 East Street, Alresford,
Hampshire SO24 9EE, UK
office@jhpbooks.com
www.johnhuntpublishing.com
www.iff-books.com

For distributor details and how to order please visit the 'Ordering' section on our website.

Text copyright: Stephen Palmer 2023

ISBN: 978 1 80341 466 9
978 1 80341 467 6 (ebook)
Library of Congress Control Number: 2022951270

A CIP catalogue record for this book is available from the British Library.

Design: Lapiz Digital Services

UK: Printed and bound by CPI Group (UK) Ltd, Croydon, CR0 4YY
Printed in North America by CPI GPS partners

We operate a distinctive and ethical publishing philosophy in
all areas of our business, from our global network of authors to
production and worldwide distribution.

Contents

To Roy Major and the good folk of Pipe Aston.

Preface

This is a book about the continuity of an idea in human culture – but not just historic culture, written down. Some ideas go way back into prehistory. Much about hunter-gatherer life in the Ice Age we know almost nothing about, since culture at the time was oral, but archaeology has given us a lot to ponder, and we can make a few educated guesses. From before one hundred thousand years ago after all, human beings were identical to those you and I know now. It was just that their circumstances were different. Yet they, like us, grappled with what they could not comprehend, or, at least, found hard to understand: death, birth, their own vivid emotions, the stars, the wandering planets, the Sun and the Moon. They also understood natural history to a degree we find hard to grasp, and therefore imagined animals in a completely different way to us. Animals were their siblings. I wrote this book in order to view human beings from the perspective of the great aurochs bull of the European plains, related to the sacred bull of Mediterranean lands. To hunt an aurochs in the Ice Age was to change the balance of the world, to reset relationships amongst the great web of living creatures. It is difficult now to imagine how that might be, but it is instructive to try.

S.P.

Introduction

When I turned sixty my partner took me for a week's holiday to Devon, and there we celebrated the occasion with visits to spectacular landscapes, with fish and chips beside the sea, and with friends old and new. As she and I wandered the historic town of Crediton we happened across a community bookshop, in which I spotted a new book by Jo Marchant called *The Human Cosmos*. This was an author I knew and liked from her book detailing the Antikythera Mechanism, so I bought the new one without hesitation.

On the very first page I read something remarkable, something which during decades of reading across archaeology, cave paintings, anthropology and prehistory I had not encountered before. This passage explained the extraordinarily close similarity between one of Lascaux's painted aurochs – the so-called Bull No. 18 – and the constellation Taurus. Now, I normally run a mile from hypotheses like this, and I've been single-minded in my opposition to ludicrous conspiracies peddled by credulous authors and amplified by the Internet. Jo Marchant's words, however, felt different. Not only was the hypothesis presented by a respected science author, the facts supporting it seemed too plain to ignore. Thinking about it, and reading other research (of which there is little owing to the opprobrium carried by suggestions like this), I was persuaded that it was reasonable to suppose that around twenty thousand years ago human beings in the deepest nadir of the Ice Age gazed at the sky and saw the shape of an aurochs bull there. They would not have called it Taurus, but they could easily have imagined it as a bull with horns and a red eye.

This was the spark that lit the tinder in my mind. From that single passage this book emerged.

By coincidence, I next read a quite brilliant work called *Mythos and Cosmos* by John Knight Lundwall. Wanting to read a book about the differences between oral and literate cultures, I checked it out online, read some great reviews, then purchased it. It turned out to be a revelation – erudite, inspired, compelling and fascinating. Yet the synchrony with Marchant's book was remarkable, since Lundwall's explained much about the way the night sky, not least that most ancient constellation we call Taurus, was used by preliterate human beings to support their structures of myth. As I read *Mythos and Cosmos* and combined it in my mind with *The Human Cosmos*, I grasped the structure of the book you now hold in your hands. I imagined a work called *I Am Taurus*, in which I would explore how human beings have viewed themselves and their place in the universe through the eyes of that glorious, celestial horned bull.

Two other books were critical in putting my own work together, both of them read long before *I Am Taurus* came into being. They were written by one of the most brilliant anthropologists of his generation, a man known across the world for his contribution to the field of the understanding of cave paintings. *The Mind in the Cave* and *Inside the Neolithic Mind* serve many purposes, but their author David Lewis-Williams' main thesis is that the structure and mode of operation of the human brain leads to certain universal images and patterns, many of which can be recognised in cave art, not just in Europe but across the world. In addition, both books put forward the enticing and plausible hypothesis that prehistoric human beings, especially via European cave art, conceived of cavern walls and later the walls of their own buildings both as residences and liminal zones through which immaterial beings (including themselves via shamanic practices) could pass.

I am also grateful for the work of Nicholas Humphrey, Jean Clottes, Karen Armstrong and Steven Mithen, all of whom have written inspirational books which tangentially bump into this one.

In this work I have synthesised a lot of my reading and thinking over the last thirty-five years, some of which I mention above. I have stayed with the facts as they are known or generally accepted as far as possible, but, inevitably, and unsurprisingly in an impressionistic book like this one, I have speculated, imagined and coloured a little. The eyes of Taurus see many things.

Chapter 1

Lascaux

They approach me with stealth, crouching down because they do not wish to be seen amidst the undergrowth, flint-tipped spears in their hands, looking for signs of weakness. But I am Taurus, and I am strong.

I am the night sky aurochs bull, my eye a star shining red and baleful, bright behind two great horns, above my shoulder six stars grouped in a tight cluster. On my cheek is another star cluster shaped as a V. I sit on the celestial plane through which the planets travel, at a cosmic crossroads where the galactic plane called the Milky Way meets that faint radiance called the Zodiacal Light. My location is a gate into another, higher realm, a meeting place, a position of significance in night skies for all in the world who view me. I am outlined by stars, a pattern obvious to all from which a more numinous symbol can be extracted: the aurochs bull of the heavens.

They approach me in Lascaux caves on their hands and knees, one carrying a stave atop which tarred plant debris has been wound, set alight from the campfire, which itself was lit from a spark glowing in kindling held inside a hollowed out hoof. Another carries a lamp made from sandstone with a carved hollow at one end, in which tallow burns, the wick made from moss. Lamps and staves are all the illumination they have, and through their eyes it makes shadows shiver, cave walls ripple, roofs and floors disappear into depthless darkness. The caves are cold, but they wear fur garments stitched using needles made from bones, nettle cords their thread.

They know these caves because they and their ancestors have been coming here for millennia. They are bound to the land in which Lascaux sits, intimate with it, knowing it in all its seasons, alert to every sign and natural countenance, which tell them where plants and herbs may be found, where water is, and where the aurochs which they hunt might be located.

To them, everything they experience in their lives is special. Though they grasp that some aspects of their lives are dull or ordinary, every other experience – the overwhelming majority – is sacred. A boulder is not just stone; that is too pragmatic an attitude. A stone has another realm behind it, an *essence*, they believe, which allows them to grasp and mentally manipulate the complexities of the world. They live in a world bright with meaning. A tree for instance is not just a tree, it is a symbol of the vitality of life. I am not just an aurochs, I am a creature of fable, metaphysical yet warm with blood also. Thus the symbols of their tales and what lies behind them are one and the same.

In their lives an opposition exists: light and dark. The sun gives bright light during daytime, but the nights are as black as pitch – deep, impenetrable, ineffable. This opposition symbolises to them one of the profound verities of their lives, which they echo when they enter Lascaux. Outside, daily life continues underneath the sun: inside, darkness dominates, a spirit world underground which they must interact with. To them, a cave is not just a hollow, it is numinous with meaning, formed because of purpose, so that it must be occupied and used with ritual reverence. Music and dance constitute much of that ritual, with reverence felt by every member of the community. Yet deeper still, at the ends of narrow passages and sudden drops, lie places meant for certain individuals alone, the shamans, set apart from their kin by virtue of their skill and experience when dealing with the ethereal realm. The people living around Lascaux are one community, tight-knit, but they are divided by character, with some accorded more status because of their otherworldly abilities. Those permitted to enter the deepest caves are few in number.

I observe shamans of Lascaux as they move around the cave system. They experience altered states of consciousness, which they characterise in ways meaningful to them and their kin. Usually such states are generated by repetitive actions, such as drumming, chanting and dancing. When experiencing such a state the mind of a shaman perceives visual, aural and haptic illusions, which to them constitute the particulars of an alternate realm. Indeed, shamans are those with access to such a realm, via what they experience as their disembodied minds travelling to and from such places. But it is their brains and nervous systems acting in altered states which create the illusion of dissociation and the disembodied journey.

The goal of such shamanic activities is to acquire wisdom from the entities of the ethereal realm, to heal, to bring rain, and to acquire information about animal herds and their whereabouts. Knowledge from shamanic journeys they believe to be critical to the survival and continuation of the community. When a shaman journeys to the ethereal realm the community says that he dies, only to live again when he returns to his body. When his body dies, he travels to exactly the same realm but never returns.

They hunt me and other animals, eating a variety of foods, with meat prized because of its nutritional value. So they hunt aurochs; they hunt *me*. They carry spears made of wood to which pointed flint heads have been attached using plant cords. Some carry narrow slings made of hide, a long loop of leather into which the spear base fits snugly, giving those hunters a longer effective arm from which the spear gains greater momentum. Such spear-throwers are dangerous to me and my kin, the throwing technique passed down from hunter to hunter, an oral wisdom held in small communities for countless generations. And they know all the natural signs, these hunters: the migration routes, the messages of hoof prints by pools, the tales told by piles of dung. They can smell my odour on the wind, can hear my calls across plains and river valleys, can see the clouds of dust thrown up by my stamping hooves, and they can see the pecking carrion birds which follow me and my kin. They are full of lore.

In Lascaux caves they approach me carrying stone vessels filled with black dyes made from manganese dioxide and charcoal, a red ochre dye made from iron oxide, and kaolin clay. A cave wall is not blank to them, however, it already contains interior structure, manifested in every bump, curve and hollow of its surface. In altered mental state they study what they can see in flambeau light, their minds alert to visual metaphors, to tricks played by dancing shadows given life by orange light. To them, cave walls are not solid, made only of stone, they are composed of other, less substantial materials, some of which represent another world. That world is a realm of the nonliving – the preborn or the deceased. It is a world of intangible essences, both human and animal. Such a realm exists *behind* the cave walls. I am on their minds because they hunt me.

Inside the hall in which my image predominates two converging rows of aurochs lie to the left and to the right. Amongst the aurochs bulls to the left are painted horses, alongside stags with antlers rendered as abnormally complex curlicues. Also to the left is an animal with a distended belly and two long, straight horns, with oval shapes painted upon its body. To the right, aurochs bulls dominate. The curves, ridges and folds of the walls form parts of these paintings, connecting envisioned animals behind the rock surface with the painted images themselves. The hindquarters of one of these bulls is delineated by a horizontal ridge of rock, and when viewed in the light of a lamp held to one side shows the outline of its rump and flank with marvellous artistry.

To those who come to see me in flickering lamplight it feels as though they are standing amongst a procession of great animals. This is indeed a communal space, not far from the surface where sunlight shines; some outside light gleams down here, though not enough by which to paint. My image is not created by a lone individual, it emerges from the efforts of many. The upper reaches of the hall are too high even for the tallest person to reach, so they build scaffolds of wood, holding them firm with twine made of twisted plant fibres. Nor are the images born of the free-floating visions of one painter, they are communally created, the overall composition considered long before work begins, with the most skilled painters undertaking each image.

Many names are attached to me: prey, hoofed beast, horned one. Many names are used to describe what lies inside my body: meat, sinews, bone and marrow, liver, heart, guts, brain. They consume or use every part of my body, with not a scrap wasted. They eat cooked meat because roasting aids their weak digestion. My hide is turned into leather, my sinews into cords, my bone into needles and other fine tools, while my marrow is eaten with delight. They consume everything which lies inside me, knowing what nutrition lies in those warm, wet, slick organs.

Yet when they approach me upon the plain or in the river valley they are ambivalent about hunting me. Every member of their community understands that the others are like themselves, each one a unique individual. Everyone understands what it means when a member of their community weeps, because they can enquire of their own personal state in order to grasp what circumstances and feelings lead to grief. Everyone realises what it means when a member of their community drools at the sight of food, because they can ask themselves what situations and bodily sensations lead to hunger. Everyone grasps that when in winter their kin shiver, they are cold, because they also have been cold in the ice and snow. They are conscious because they empathise, using themselves as exemplars to understand the behaviour of others in their community.

But that mode of life they also apply to me. I am an animal who eats grass and plants; other animals eat meat, like the cave lion or the hunting eagle. I eat grass because I am lived *by* my life, told what to do and when to do it by instinct alone. I am Taurus, and I do not think as they do. But those hunters on the plain do think, and they believe me to be exactly like them: possessed of individuality, of an essence, of a thinking mind. They do not see just aurochs drinking water at the bottom of the riverbank, they see one such as themselves, alert to the wonders and perils of the world, and making wondrous stories from aquatic realms. So when they hunt me and kill me with spears they feel ambivalent, knowing that in taking my life they are changing the balance of the world. They believe all animals are siblings akin to themselves, and so they feel guilt when I die. Animals stand alongside human beings *in* the natural world, entangled with it, bound to it, forever a tiny part of its complex, difficult, perilous, yet meaningful state. So they hunt and kill not just physically but with meaning, by persuading the animal to surrender itself or by asking the permission of its master; and what they take must in some sense be put back afterwards.

A hunt is not mere functional activity. It is profound, and it has meaning for them, one which they manifest and celebrate in song, dance, and in art.

When they approach the Lascaux hall where so many aurochs bulls are painted, they carry in their hands the implements necessary for fixing images of me and my kin upon stone. Although those bulls have been visualised before, in trances, in dreams and in visions, they are manifesting bulls they believe already exist inside cave walls. They notice in accidental rock forms a bull's head, the distinctive shape of its back, a haunch or a flicking tail, and from such starting points they complete the image in black and red. There is never any natural context for such paintings, no sky or ground or long grass. The bulls float in speleological space, often with their hooves unpainted, or rendered as if relaxed. When I see such an image I feel as though I am floating in some ethereal realm.

When they paint they do not use only their hands and fingers to render my image. Charcoal powder mixed with saliva sticks well to cave walls covered with glistening white calcite. They take paint into their mouths and blow it with subtle brilliance upon the walls, rendering hide, and features of the head and face. But although when they paint me in Lascaux caves they are fixing me, it is not just the final image which matters – it is the *act* of painting. To blow paint upon a permeable wall of the intangible realm is an act of breathing made using their most intimate and precious resource: their self. Breath symbolises that self, and when they paint by blowing they connect themselves in the deepest way possible to the entities within the cave wall. One of the aurochs in the hall has its nose created by blowing paint, while another has a line at its mouth signifying breath. Breathing is connecting.

That connection, however, is not one-way, for the painters are creatures of metaphor, understanding that mental concepts may be used as a mode of representation. Though in painting they fix vision images of the aurochs they hunt, thereby manifesting those animals, they also bring bones into the caves, forcing fragments into crevices in an act which is the reverse of painting. In this way the traffic between the tangible land upon which they live and the intangible one just behind cave walls is maintained, supporting the balance of life in circumstances which admit both of life and of death.

They approach me in night skies through the stories kept in their minds. Such stories are many things: explanations of the world around them, which they sculpt into meaning, oral histories of themselves and their community, which they pass on down the generations that nothing about their lives and the land upon which they live is lost, and guides for behaviour, which they relate using themselves as characters.

In community stories there are heroes and there are tricksters. Heroes are necessary because such larger-than-life characters are required by the limitations and onerous character of oral memory. They who gaze up at me at night must strip their myths of unnecessary detail, paring back to the essential parts of the tale, that those parts never be forgotten in the telling; for the only memories available are human ones. A hero is at once human and more than human, a cipher, a metaphor, yet also a feeling, thinking person with a family and kin. A hero is memorable, indeed, unforgettable. A hero appears more than human because the deeds of such individuals are easily remembered when stitched together into a story told around a campfire. Nobody forgets the one who journeyed beyond the veil of night. Nobody forgets they who sailed amongst the stars in search of wisdom. Nobody forgets the person who dared penetrate the dense, silent, crepuscular regions of the afterlife.

Tricksters are necessary because the human beings who see me upon the plain in aurochs bull guise and in the night sky see others of their kin in a similar light. They have bodies, and they have another, ineffable part: individuality, character, which some call spirit.

Each member of the community understands that everyone in that community has a mind just like their own; that is, everybody has a theory of mind. This situation arises because the behaviour of their family and their kin is so complex and difficult to judge the only way to manage it when living in a community is to use themselves as an exemplar. Their minds are teeming with thoughts and questions about their kin – what their desires and needs are, what their thoughts might be, what they are planning. Now, the aurochs has a brown eye; it does not need to know where one of the herd might be directing its gaze. But human beings have eyes mostly white, which makes knowing the direction in which they are looking much easier to guess. In this, and in innumerable other ways, everyone in the community has a grasp of the state and quality of another individual's mind – their theory of such minds. The trickster is the metaphorical embodiment of this state of affairs. Tricksters know the hearts and minds of others in their community. They are the ones with lore, who weave knowing into knowledge and knowledge into wisdom. Better than any other, a trickster is one with insight into how their kin think, and therefore how they behave. As I look down upon them from my celestial place I see the value and potency of such understanding.

Their understanding is founded upon similarity. When at night they look up at me delineated by stars they imagine other forms most similar, including the forms painted upon Lascaux cave walls. Their mode of thought is metaphor, which is the representation of things by other things. Throughout their lives they compare one thing with another: the fissure in a cave wall with the back of an aurochs, the fold of rock with a haunch. They compare the pattern of stars comprising me with living aurochs on the plain, using one as a metaphor for the other. In this way, alive to the innumerable similarities, morphological echoes and other patterns of their world, they build up complex mythological tales, screeds of symbolic meaning, made all the more memorable in their oral culture by techniques of memory.

Because they live by metaphor and story, they use nature as their main mental framework. Remembering many stories is difficult, so they use the natural world as a mnemonic resource. It is for this reason that the star-strewn night sky is fundamental to their lives – stars including those comprising me. I am conspicuous, rendered for all to see, my celestial pattern bold and obvious. I therefore *demand* an explanation, a metaphor and a host of stories. Seeing me in stellar form, they cannot ignore me; they have to explain, elaborate, find meaning. Such is an intrinsic part of their minds. Within cave walls they discern shapes moulded by time and water, which they use to evoke the aurochs of their dreams and visions. In night skies they discern patterns made by the brighter stars, which they use to evoke me in all my strength and majesty.

They approach time both through the animals around them and by the wheeling stars above. They understand that aurochs have an annual cycle of life, just like other animals: horses and deer for instance. Horses they paint first during the year, depicting them with the rough coats and long tails of the end of winter. Stags they paint last, with great antlers signifying the rut of autumn. I am painted second and in summer garb. All three species are painted in our mating season. Thus the greater structure of Lascaux, its *meaning*, is one of the cosmic cycles of the universe as seen from that tiny landscape around the cave system. Those cave painters are profoundly aware of the night sky, and from it they take the meaning of their lives and the world in which they live. Stars lead their way, patterns creating metaphors, which diversify and mix and sophisticate in their minds, bringing ever more complexity to their social and cultural lives.

Time for them is cyclical. When they paint me in the Lascaux hall – the largest aurochs bull, and the finest painting in the whole cave system – that group of six stars so important to them rises a little before sunrise in the middle of autumn, reaching zenith at the beginning of spring, and disappearing from view at the waning of summer. The vanishing and reappearance of that star cluster above my shoulder marks the mating season of my kin, the aurochs.

In spring I command the night sky above the hills adjacent to Lascaux, bestriding the celestial firmament, my horns obvious, my eye bright and red, a metaphor of the cosmos, of time, and of the life of those who paint me in the caves. I am great, I am strong, I am majestic.

I am Taurus.

Chapter 2

Çatalhöyük

They approach me as I stand amongst the herd, wearing clothes made of woven cloth, shoes of leather upon their feet, their stomachs filled with unleavened bread made from wheat that they plant, nurture and gather. They are looking for me, the special one, but I am Taurus and I am difficult to spot.

I am the bull set within a mud brick wall of their house, my skull taken from a dead and decaying animal, defleshed then cleaned, and transferred inside. Surrounded by mud, stones and plant debris, I exist in that liminal, numinous region between the real world and that of another realm, my horns sticking out into the room: long, pointed, and impossible to ignore. I am covered with plaster, terrestrial bull and celestial bull, an intermediary between worlds.

They approach me inside their labyrinth of Çatalhöyük houses, each one entered through an opening made in the roof. Clambering down a ladder leant against the southern wall they find themselves inside a small, gloomy chamber, illuminated by windows set high. Some rooms have an entrance shaft in the northern side, and a bench which helps them obtain access. The chambers are self-contained, most lacking access to any other room of the settlement. Towards the centre of Çatalhöyük the chambers are decorated with striking, sophisticated designs, including those featuring the skulls of my kind. But they do not consider these chambers to be rooms of one purpose; the entire settlement is approached as if the mundane and sacred are one and the same. These interior decorated chambers, however, are a little different in character from other living spaces, in that they often have small portholes in the walls leading into tiny, unlit rooms. Thus, as they move from outdoors into the settlement and from one chamber to another, they are performing a ritual at once secular and metaphysically meaningful.

In aeons past I was painted upon cave walls. They do not know me from those caverns of western Europe, but a few days' walk to the south lie mountains permeated with caves which they do know, full of stalactites and stalagmites, and veins of green and blue apatite. From the stalactites and stalagmites they knock off pieces, which they take back with them to Çatalhöyük, while with the fragments of apatite they make bead necklaces. In other regions of Anatolia, on the slopes of volcanoes, they find obsidian, the black glass which is so sharp it must be handed with great care. To them the underground world is an otherworldly place, and this they replicate in the design of their settlement. Movement around Çatalhöyük is a meaningful equivalent to moving around caves as their ancestors used to. Thus, when they move through portholes from those richly decorated chambers into the tiny chambers set behind them, they are performing a profound act akin to entering the deepest cavern, inside which, in epochs past, their ancestors experienced visions.

The interior chambers are decorated with images of me. Against the east wall two columns are set near their respective corners, each with a protrusion at the bottom and the top. Between these columns lies a panel into which three large bull heads have been set, their horns projecting upwards and out. Below these heads lie many much smaller bull head images, arranged in two layers. To the right of the right-hand column lies another, smaller bull head, while to the left of the left-hand column two small bull heads are placed. Upon the ground before the panel a low barrier has been built with two raised projections on each side, into which bull horns have been set, in such a way that the horns project into the space between wall and barrier. To the right of the right-hand pair of horns is a square hearth. The north wall is plastered, and into that overlay an image of me has been cut, my head facing away from the wall, my belly full, my legs short and stubby.

I am *everywhere* in this room. Its odour is of stale sweat and straw tainted with animal dung. They see me, and they smell me.

They see me in herds on the middle level of their three-tiered cosmos. Above them, the sky: below them the underworld. They are predisposed to view the world in such a way because of these circumstances in which they live, and because of the storytelling gifts given to them by their conscious minds. Connecting the three levels is the central column of their imagined cosmos, the pillar of the world. This arrangement of triple tier and connecting column is the basis for all that they imagine of the realms to which they have access, yet without an intermediary they are voiceless and impotent, unless one of them through shamanic means can *become* that intermediary, and journey between realms.

I am mighty amongst them. I am strong, massive, and with my horns I can stab, and kill. They need me for so much: for meat, which they consume, for hide, which they work into a multitude of forms, for horn, which they carve, and for my inner organs, which, being packed with nutrients, they eat with gladness. My marrow is theirs to suck out of every bone; my hairs are theirs to bind into brushes.

I am their intermediary. Inside those tiny unlit chambers the skulls of bulls cluster around the bottom of their metaphorical world pillar, symbolising my ability to accompany them on mental journeys, to be a guide, even to act as a vehicle should they be able to transform themselves into me. I live in a liminal location: the wall. Such pillars in Çatalhöyük are posts made from the wood of oak and juniper trees, but they do nothing to support the mud brick structure – they are symbolic, the juniper brought from those mountains to the south in which the apatite caves lie. Some posts are covered with red ochre, indicating metaphysical importance, and one has a zigzag pattern painted upon it. In such ways they emphasise to themselves the importance of their world pillar.

In entering the settlement from the outside or leaving it to go elsewhere they are undertaking symbolically meaningful acts, clambering up and down ladders which to shamans are the bridges between worlds. Such simple deeds are to them deeply meaningful: evocations of the vertical plane, personal expressions of the character of their world. Mundane and transcendental are merged into one lived whole.

They conceive of the underworld *inside* Çatalhöyük. Theirs is a settlement imbued with many metaphors: female figures, felines, and of course myself. While the skulls of my kind crowd around the bases of pillars and upon them, felines never do, while the overwhelming majority of female figures are situated elsewhere. But they are aware of the transience of their existence in the middle world; they know they will one day die. Death is a profoundly puzzling experience for them, they who are so aware of themselves, of the depth and intensity of their lives, and of their inexplicable origin inside the bellies of women. Each dead ancestor represents a fragment of the remembered past, spoken around fires at night in stories recalled using mnemonic techniques honed over countless generations.

They place some of their dead kin underneath platforms set upon the floors of their houses. In Çatalhöyük many individuals lie like this. In this way they bring their concept of the underworld into their built environment, their mundane *and* metaphysical environment, which in their eyes is home yet which also encapsulates and symbolises the tripartite world.

They live in Çatalhöyük for many generations, building on top of previously constructed rooms so that the settlement over centuries and then millennia becomes a mound of many levels, with the most ancient level at the bottom and the occupied one at the top. In time, it will be as tall as ten men each standing upon the shoulders of another.

From inside the walls of this settlement I look out from my vantage point. I reside in a constructed liminal zone, peering at them, ready to be their guide. In acts of careful devotion they replaster each of my heads time and time again, ensuring that the complex meaning of face and horns is not lost. Later, when Çatalhöyük is a great mound upon the plain, they embed real skulls from my kin into their walls, each of which they cover with a thin veneer of plaster. This act of replastering is as important as my final image, a visage which hardly changes, therefore retaining its symbolic power over generations.

Çatalhöyük walls are a transcendent medium with a shamanic role into which they place some objects unrelated to me: the teeth of predatory mammals, the beaks of local vultures, the jaws of beasts they hunt, and the skulls of small creatures. But one wall contains something special. It is the tiny skeleton of a stillborn child, wrapped in woven cloth and placed with a shard of obsidian and a piece of shell into a mud brick, which then is used in house construction.

So motion through a wall is in two directions, as their Ice Age ancestors believed when inside cave systems. They who live in the settlement imagine that spirit creatures can emerge, as depicted by their bull skulls, but they themselves do possess the ability to pass within the walls through ritual. Many walls contain niches painted with red ochre. Just as their ancestors placed bone fragments into cave wall cracks, they place objects into these niches.

I am Taurus, the special one. A shaman can become me when on a journey, assuming my form and all that goes with it: the horns, the bulk, and feelings inside too, for a shaman must be open to all the experiences I feel when I am alive. I am the intermediary between the three levels of the cosmos. I am strong, knowledgeable and clever.

Some shamans are called to their ability by dreams and visions, which emerge like spring water bubbling up from the unknown depths of the earth. In their minds they see things not visible to those more ordinary; this is the water of experience. Their subconscious minds are where all such trance sensations begin; this is the earth of origin. Water from the ground, thought and feeling from the deeper mind. It has always been this way for them.

Because of their perception of intangible realms, shamans are reborn after they are born of woman. To become a shaman is to be newly created. In Çatalhöyük houses they make sculptures of women giving birth to me, horns and all, metaphorically portraying the manner of shaman rebirth and the authenticity of their transformation into me. I emerge from the depths of shamanic minds: I emerge from the depths of women's bodies. These are one and the same thing.

A shaman is not depicting me, or playing at being me – they *are* me. The identity is exact and whole.

They believe their ancestors are all around Çatalhöyük. But those ancestors during the Ice Age lived in caves, gathering food and hunting it, living in small communities entirely dependent upon the ways of the land. That land always provided for them, but nevertheless they did depend upon it. In Çatalhöyük, circumstances are different. Theirs is a constructed environment, based on the lore and symbolism of the cave. With mud bricks they build houses all squeezed against one another, making artificial caves that suit them well. These are safe places, protected from the vagaries of bad weather and even from attack by wild animals. Yet in creating Çatalhöyük they have introduced a category into their social lives never known before, manifested through a novel opposition – wild: civilised. They experience an opposition between the natural world and their social world.

Amongst local grasses they notice one type whose stems do not shatter when the ears of grain are tawny ripe. In other species seeds are scattered by wind and animals once the stems have broken, but this new variety does not follow such a path, meaning it is available for harvest once the ears are ripe. Thus in Çatalhöyük they take advantage of a random event in the natural world, keeping some of the seed to throw upon the fields later, that during the following year more ears of grain ripen to feed them. In this way they domesticate the grass from which they bake their unleavened bread.

By chance, some of the species of animals living around them are malleable herd creatures – a broader selection of such creatures than anywhere else in the world. I am one of those species, from whom cattle descend, and there are also sheep, akin to the mountain mouflon, and pigs descended from wild boar. Goats, akin to the mountain ibex, are not quite so fond of living in herds. These four species can all be domesticated.

I am Taurus, the special one, but I am difficult to spot when in my herd. Yet for the community's shaman such identification is easier, and this is deemed proof of their great power, verifying that they deserve high esteem. Much of this prestige is due to their ability to interact with the intangible world where I live. Shamanism therefore is a socially divisive force, allowing a small elite to enjoy the benefits of high status. In order to manipulate the herds of aurochs, shamans perform various tasks, sometimes guiding prey animals into hunters' traps, occasionally persuading a high status animal to give up one of their kin, sometimes acting to ensure the continued reproduction of the animals. In such ways, over time, a metaphor is created, the metaphor of human control of animals in herds. They do this not only for food and animal products, but also for sacrifice and the continuation of natural balance. Thus the cultural metaphor of shamanic influence migrates into the real world, allowing shamans a certain amount of presumed control over actual aurochs herds. This is the beginning of domestication.

The accidental consequence of domestication becoming more sophisticated over millennia is a new relationship with land and nature. Animals can now be placed into two categories, one wild – entirely of nature – and one domesticated, that is, belonging to civilisation. My kin the aurochs in their great herds are important to those living in Çatalhöyük, though not in the end as important as other herd animals. But domestication leads to a new way of gathering, and therefore of storing, food. Domestication changes the odds of life.

The plain upon which Çatalhöyük is built overflows with wild animals, including pigs descended from boar and a species of deer. But when some of those species, already prone to being corralled because of their instinctive herding character, become quiescent, the shamans often turn to larger, stronger, more ferocious and dangerous animals. I stand amongst that group, as do bears and leopards – the latter a common feature of Çatalhöyük sculptures. In this way, my imagery dominates the cultural landscape of those who live in that settlement.

Çatalhöyük does not last forever. It fades, it fails. Over long periods of time the social prestige of shamans who become me and control herds of aurochs diminishes, not least because of the success of domestication on a wider scale. By the end of the settlement, many other lands and settlements own herds of domesticated animals, from which undoubted material benefits come. Yet those benefits emerged from the bull shamans of Çatalhöyük.

In darkened chambers my head pokes out from plastered panels set between columns. I am a multitude here, a cosmic herd, looking out from that fluid zone composed simultaneously of mud brick and intangible substances; decorated walls, special, ethereal, which mark the boundary between worlds. I am the aurochs bull of the liminal zone, my breath a snort, my eye red fierce, my horns a great crescent echoing the fingernail moon just after new. I am a guide, I am cunning, and I make journeys between three realms, my hooves sure and steady on the juniper pole connecting the tripartite cosmos. I am confident, I am wise, I am strong.

I am Taurus.

Chapter 3

Alacahöyük

They sacrifice me using sharp metal weapons, spilling my lifeblood upon the ground. I am a sacred animal to them, a bull who has survived from the time of ice and freezing winds, tough, magnificent, walking with confidence upon Anatolian lands, and further south into the Levant. But I am fierce too, and dangerous.

In statuette form I am manufactured, my horns projecting upwards from the sides of my head, their pointed ends curving backwards a little, sturdy on four feet, my head displaying pride and surety of character. I look forwards without hesitation or uncertainty. I am poised and assured, knowing what I can do should the occasion demand. I am strong and I am majestic.

The people known as the Hattians revere me, their sacred standards showing me in all my glory at the settlement called Alacahöyük. I am their sacred bull, surviving in later stories for the Hittite and Hurrian peoples as two bulls named Day and Night, Seri and Hurri, who upon their broad backs carry the weather deity Teshup. Sometimes we pull the chariot in which he stands as we graze upon demolished cities.

Teshup is the deity of storms and the awesome, unreachable, imperishable blue skies. The empyrean is a metaphor for these people in all its grandeur, a statement of transcendence, of permanence, a realm utterly different from the muddy, messy, terrestrial world. For them Teshup is vast, unstoppable, and full of violent force. He can appear at will to lash the ground with flood and lightning, bellowing with thunder, striking with hail and winds. They always feel defenceless before him, at the mercy of his meteorological whim, exposed to his fury. This deity of infinite blue skies comes and goes across Anatolia as he pleases, and I pull him according to his wishes. He carries three thunderbolts and a double-headed axe. But I too am strong and worthy of awe, so I fear no deity of the storm. Indeed, the crown Teshup wears has two horns signifying that he and I are akin, that we complement one another, that we are united. They who live in Alacahöyük know that fact and do not forget it.

Teshup, Arinnitti, Nerik their son, Kurunda the hunter and Hannahanna the mother goddess each follow their own life, as do all Hittite deities, but for sustenance they require the reverence of human worshippers. Likewise, the people of Alacahöyük need the attention of their deities in order to survive. Thus a mutually beneficial culture appears in which deities live inside temples, where they find food and drink, cleansing, and even diversions and amusements. Yet they never stay long inside their temples, journeying across land and sky in pursuit of their own destinies. These deities feel many moods and can be angered, whereupon they have to be tempted with urgent prayers to return, else the lands of Anatolia sicken and die. Yet if a devotee is honest about their delinquency they are not held to account by the deity, and all is returned to normal.

These people know the depth of Teshup's antiquity, recorded in their oral myths so it is never forgotten. Originally named Taru by the Hattians, he acquires over the centuries various similar names all deriving from the word *tarh*, meaning to defeat and conquer. His grandfather is Anu, who has his manhood bitten off by his son Kumarbi, a deed through which Teshup is conceived and born. Aeons later another culture will tell the same story using the names Uranus, Cronus and Zeus. Yet this is often the way with tales of creation spoken by these families of related peoples, linked by land, by language, and by ever-developing culture. Uranus, manifestation of imperishable empyrean, is Anu; Cronus, overthrower and castrator, is Kumarbi; Zeus with his thunderbolt is Teshup.

All these myths have changed since the days of hunting and gathering. Myths are in large part oral guides to behaviour, so when circumstances lead to changes in social structure, myths are altered to match. Yet the eternal truths of life – death, birth, sex, food and water, struggle and cooperation – remain at the heart of myths. As wide-eyed wonder at the vast blue sky recedes, as agriculture takes over from hunting and gathering, so those most ancient sky deities, who were the first divine anthropomorphisms, fade into the background, to wither, to be eaten or unmanned, to be fought and killed.

Alacahöyük is a vibrant settlement even before the Hittites appear in Anatolia. The royal tombs there are full of ritual objects and are decorated with complex designs. One tomb features pairs of bull skulls and foot bones. Lying less than a day's journey from the Hittite capital city of Hattusa, Alacahöyük is a cult centre for the sun goddess Arinnitti, after which the original name of the settlement, Arinna, is taken. Arinnitti is the wife of Tarhun, that is, Teshup. Sun discs are the symbol of her cult, and she is deemed an exalted patron of the state.

Many stone structures characterise this settlement. They carve sphinxes here, creatures with the bodies of lions and human heads, notably at the southern gate – the Sphinx Gate. There is conflict in this part of the world, so such gates have inner and outer ways, ramps leading up for archers to use. The Sphinx Gate itself has a pair of stone guardians and is built between two monumental towers. The stones they use to build walls are massive, but each fits perfectly with those around it, yet it is the two massive sphinxes which induce the most awe, staring down at visitors from on high. The settlement is also decorated with standing slabs, each carved with intricate designs showing people and beasts of the field, with one showing me, my head lowered, horns ready for attack.

They are dextrous and creative in their use of clay, upon which they depend for so much. Clay in essence is a fine-grained soil impregnated with various clay minerals. When wet, clay acquires plasticity owing to the film of water surrounding each particle. They burn wood inside a kiln and place the objects to be fired inside, the heat of the fire making the clay hard and brittle, removing its plasticity forever. In this way, and with ingenuity, the people of Alacahöyük make a wide range of ceramics.

In their early royal tombs they lay many bronze objects. These fall into three main groups: sun discs sacred to Arinnitti, figurines of stags, which are a sacred animal alongside me, and figurines based on my own form. The sun discs are complex, each showing multiple interior patterns pierced through – spoked, diamond-shaped, swastika, triangular and square. Some of these discs have projections around the outside with further patterning. Most are asymmetric, but some are symmetric.

The stags all have elaborate antlers, many exaggerated for metaphorical effect. The image style is graceful, with stags depicted as slender and athletic – a stylised, not a naturalistic rendering. In most cases they are mounted on quadripartite stands, often with all four legs brought together as if they are about to leap. Sometimes two stags are shown side by side.

I am also depicted with exaggerated grace and athleticism, my body slim, usually with all four legs close together in the stag pose. My horns are long, curved backwards, with very sharp tips.

The metal they call bronze is an alloy of copper and tin, the former metal being present to around nine parts out of ten. This tin is added to ensure the metal remains liquid for longer than copper alone, making for a better casting technique and more durable figurines. Tin, however, is only found in certain locations; they know it from its ore cassiterite, which they smelt separately from the copper. Some bronze is made with arsenic; other trace metals may be added.

Before bronze is devised they make objects of many sizes in stone alone, but a thousand years or so before they begin writing they are already using a technique called lost wax casting with this metal. Bronze has a long history.

A sculptor using the lost wax technique first makes a model in wax of the figurine they wish to cast, sculpting all the necessary details as if it is the finished object. This model is then surrounded by moist, malleable clay, which is baked in order to melt the wax and harden the clay into a mould. It is this mould which receives the molten bronze during the casting of the figurine. When the metal is cool the clay is smashed off, revealing the figurine, which can then be tidied up and prepared for any additional decorative metals – silver or gold for instance.

All the Alacahöyük objects, and others, are offered as grave gifts in royal tombs. Some are carried in ritual processions and some are attached to chariots. The quadripartite bronze stands merge to a single bar attached to a long wooden pole, by which an object can be carried with due reverence. Bronze, however, is not the only metal utilised by these very early smiths, since they make use of silver for inlays, and embellish antlers and noses with silver or gold. The sun discs are usually round, but can be semicircular or diamond-shaped, their interiors pierced through into grilles, many of which are further decorated with rays or lines of metal. Often there are animals depicted too: myself, and the stags. Other animals of note appear, such as lions, which they deem a special kind of beast, though not of my own status. The sun discs often have stylised horns appended, and all have pegs at the bottom where the wooden pole is attached.

Just as the most ancient of sky deities have faded, so has the connection between symbol and reality; the sun is now a pierced metal disc as well as their prime life-giver. These people wield symbols in a different way from their distant ancestors, separating corporeal form from the mythic dimension. It is a hint of what is to come. No more do the people of Alacahöyük wander in a sacred world, a golden age as they conceive it, and they feel much less direct contact with the transcendent world. Mundanity has begun to enfold them, to corrode them, to make them impure. Now they tell stories of their first world, that golden age so far in the past, from which they are sundered. They feel lost and grief-stricken.

Yet in that sense of isolation and exile they do have one aid. Being guides to living, their myths are a nascent form of inner enquiry – of *psychology*. These people wrestle with forces which they feel inside themselves – love, jealousy, envy, fury, grief – and they struggle to cope. In myths, however, they find a method of understanding. Through tales of great individuals in extreme times undertaking extraordinary journeys they manage to grasp those inner forces which are generated by the human condition itself, by their minds, which are not akin to animal minds at all. In doing this they are at once soothed and granted coherence, that crucial coherence which is the foundation of human meaning. And in lives of meaning they find solace and hope.

Upon a wooden pole I stand in bronze majesty, slender and athletic, my horns extended and curved backwards into sharp points. My feet rest upon four stands merging into one, which links to a pole. I am alert, majestic, an epitome of power and vitality. I am worthy of reverence. In hard, resonant, dark metal I am carried high at the front of ritual processions, that my strength and potency never be ignored or forgotten.

I am Taurus.

Chapter 4

Uruk

They approach me inside their temple, whose dimensions are of the cube and whose cultic volume represents their construction of the cosmos. All around their city they build ziggurats, also cubic, whose seven levels represent seven cosmic spheres. Each of these spheres is the location for a heavenly body in motion: the Sun, the Moon, the innermost planet, the morning or evening star, the red planet, the bright planet and the furthermost planet. Like their ancestors they wish to take the cosmic order and introduce it into the world where they live, remaking it as their built environment. Inside their temple the air smells of sandalwood and frankincense. In sacred images my horns are shaped as a crescent, which I use to push the Moon across the sky; for when the Moon is a fingernail crescent an ashen ghost of the full disc can be seen. This is a busy place, bustling with activity, and in it I am revered.

I am the Great Bull of Heaven, arrayed in stars upon the night sky, associated with the goddess Inanna and linked to the morning star. I am ancient, remembered over millennia from the time when ice covered more of the world, my horns and eye visible for all to see. I am depicted on clay cylinder seals, as a bull, also as a winged gate. I watch, and I see.

I look down upon them and observe the city of Uruk. They have a new form of communication, made using wedge-shaped marks on tablets of damp clay: cuneiform. This mode of communication has existed for five hundred years, and it is beginning to transform their society. For previous generations theirs has been an oral culture, knowledge and much more passed down in the form of spoken myths, which using mnemonic techniques has kept everything they know at the heart of their culture. But now their great store of knowledge is being transferred from dynamic minds filled with wonder at the world to clay tablets, which, although possessed of some advantages, are eroding their myths and rendering them redundant.

As I peer down from the heavens I see the king of Uruk known as Gilgamesh. His father was Lugalbanda and his mother Ninsun, the cow goddess, related to me through her bovine character. Gilgamesh is two-thirds deity and one-third human, and when they write his name on clay tablets they use the divine determinative to show that he is more than human. He is an immense man, whose stride measures seventy-two cubits. He towers over all others and is mighty in exploits, exploring the ocean, travelling across wilderness, and perceiving the very foundations of the land. Yet this majestic perception is not of mundane places, it is the measurement of earth and heaven, making Gilgamesh wise in many arcane things. He observes cryptic circumstances and reveals them to his kin; he carries tidings from before the great flood; he establishes sacred places and returns old rites to their proper place. Uruk he builds within seven walls, aided by seven masters of wisdom. Thus he is king of the city, of the temple, and of the cult within it. He is the restorer of the sacrosanct locations of Sumer.

Yet in his manner and behaviour he is harsh. He is a tyrant. He beds women as he likes, and bullies men with unjust severity. In their anger and distress the people of Uruk petition their deities, asking for justice and peace, and in reply those deities create a man of the wild, Enkidu, who is a mortal but who is hairy, uncouth, wild and swift as a gazelle. Enkidu is a man of untamed nature, whose appearance and manner set him apart from they who live in Uruk, who write, live in houses and follow rituals and rites. Yet Enkidu in time finds the lives of those within city walls to his liking, and soon one of the sacred temple women who are wise in the arts of the bed seduces him, introducing him to the possibility of having friendships with the citizens. Much flows from this encounter. Enkidu learns how to eat like a citizen, not like a beast; he discovers how to wear clothes, not to rely on his covering of hair; he learns how to contribute to Uruk by becoming a shepherd and a watchman of the night; and he discovers the possibility of love. Yet he has lost most of what he had before, for he is rejected by animals of the wilderness with whom he was raised. Worse, in becoming civilised, he is taught to feel antagonistic towards lions and wolves. He has lost that innocence founded in nature.

One day, Enkidu hears about the despotic behaviour of Gilgamesh. Enkidu is a very strong man, and, infuriated by the tyranny of his king, he challenges Gilgamesh to a duel. The pair meet at a wedding and at once combat commences. By the gate of Uruk, Gilgamesh approaches his foe, but Enkidu pokes out his foot and trips Gilgamesh, whereupon the pair take upon themselves the martial mode of my kin the bull, charging and grappling with one another. Door posts splinter and the very walls of the city tremble as the wrestling pair undertake their duel, snorting in their exertion just as I do. It is as though the forces of nature are shaking the foundations of civilisation, Enkidu versus Gilgamesh, with me the intermediary. Yet in the end Gilgamesh proves too strong for his opponent, and he casts Enkidu down for the final time. The two men, however, now feel respect for one another, and after a while become firm friends, convinced that each is somehow a version of the other, that they are kin even, albeit that at first they formed an opposition of nature and culture.

Both Gilgamesh and Enkidu feel the need to gain fame and fortune, so after further discussion they leave Uruk and travel to the Cedar Forest, where, despite warnings about the peril they will find, they decide to attack Humbaba, who is the great ogre of the place, protecting every cedar tree. In this venture the pair enjoy the assistance of Shamash, who is the sun god, and his wife Aya. Eventually, with Humbaba overcome inside a sacred grove of the goddess Inanna, the pair enjoy the fruits of their venture, felling trees at their whim and fancy. Enkidu chooses the finest, mightiest cedar, and from it fashions a door for the temple of Enlil, the god of winds, earth and universal air.

Having returned to Uruk, however, Gilgamesh finds himself in a dangerous situation. Inanna approaches him, hoping to entice him into her bed, but he rejects her. Angered by this deed, spurned and feeling abandoned, Inanna calls upon her father Anu, asking his permission to take me, the Great Bull of Heaven, down to the earthly realm. Anu agrees to this request, placing in her hand a rope with which to lead me, and in this way Inanna and I depart the heavens. So it is that I journey down from my world to the world in which Uruk stands, there to do battle.

The middle world of the tripartite cosmos is not at all like my world. I find forests and woods, and rivers flowing bright and silvery down to the sea. I see mountains, and great plains around them. I see two rivers flowing in parallel from high peaks across the lands of Sumer, sometimes distant from each other, sometimes close, flowing down to a great delta, low-lying and flooded, where marshes lie and the airs are malodorous and humid. There are deserts and plateaux, hills and valleys.

I am the Great Bull of Heaven who is filled with fury. The lands of Sumer are mine to lay waste to, and this I do with a zeal born of rage. I swallow entire rivers and open deadly pits out of bare ground. I kill and consume any warrior standing in my way. I am massive, a force nobody dares halt, he who makes earthquakes with his hooves, he who snorts out great winds from his nostrils to make cedars crash to the ground, he whose weight and power is manifested in thick bone and hard, bunched muscle. I gnash and I groan, wielding my horns with wild abandon as I bestride Sumer and make all who see me wail in terror.

Now Gilgamesh and Enkidu realise they must face me. From their city they emerge to take me on in bitter combat. We clash and we wrestle, they trying to cast me to the ground and there dispatch me in the dust, me attempting to gore them, to pierce their sides, to stab a horn into their bellies and bring forth blood and guts. This struggle goes on and on, neither of us able to outwit and overcome the other. I paw the ground and glare at them with my red eye, snorting and groaning, while they crouch and spit on their hands, staring at me, sizing me up, speaking in low voices to one another.

Alas, in the end, they are too wily for me. With terrible blows they strike me to the ground, eviscerating me, then separating my head from my body. I am defeated. Gilgamesh in his joy cuts the horns from my head and takes them back in glorious triumph to his palace inside Uruk, where he places them upon a wall as a trophy of his and Enkidu's victory. Yet even that ignominy is not enough for the pair. Enkidu rips a haunch from my body and with a mighty throw casts it at Inanna herself.

The deed is done, and I am dispatched. Inanna takes what remains of my carcass and with women of her cult performs ancient rites over it. Yet the deities of Sumer are angry at all which has transpired, declaring that one of the pair must perish; and in the end it is Enkidu whose body fails him, so that he becomes sick, and dies. This distresses Gilgamesh to the extent of making him inconsolable, and in his pain he declares he will seek out the secrets of immortality itself in order to return his friend to the world of the living. That, however, relating events unconcerned with me, is another story...

They call their leader of high status a king. These men – and they almost always are men – emerge around the time people begin writing and living in cities, although the roots of that slow, insidious male domination stretch back into earlier millennia. It is a merging of divine power and secular authority which, over long periods of time, creates the position of monarch, resplendent at the top of a hierarchy composed of increasingly more important and less numerous individuals. The summit role is personified by the king. Quite often a royal individual is somehow identified with a particular deity; this person can be of low or high status, and they are not always male. There are queens in Sumer too.

Kings live in city states, which emerge before nation states. Often a king will try to usurp the position of other kings in nearby cities then unify a kingdom. These formative nation states are unstable and often disintegrate, though a wily king has various tactics he can use to stabilise his kingdom. Since he will be attempting to create a dynasty through his family, he can indoctrinate his children or use his daughters to make suitable alliance-forming marriages; and he can use violence to follow the path of military conquest. This is the era of bronze weapons and copious bloodshed. Daughters can also be installed as priestesses in the lunar temples of Inanna, known elsewhere as Ishtar. Many kings attempt to establish garrison cities, making marriages and other alliances to unify their kingdoms, though they can also appoint trusted individuals to govern otherwise troublesome city states.

A king sees himself as a military power, as a law-maker – he will pronounce edicts for his city – and as a favoured individual. He sees slaves in his city: the spoils of war. He sees women slaves in temples. Yet he does not have it all his own way. There is tension between him and other property-holding groups: private landholders, private slaveholders, the priests of temples and free communities of landholders. Usurpers with military strength are common in this period, and, to kings, war is well known.

Women thrive according to their status. High class women can hold power and even represent the king. The daughters of kings are appointed high priestess of the Moon God or of Ishtar, wearing a distinctive costume of a cap with a raised rim, a folded body garment, and ostentatious jewellery. Living within a sacred residence, notably the Eanna shrine of Uruk which is the region's main cultic centre, they take charge of temple business and perform all the usual rites. Rarely do they marry, however.

Yet it is in the city state of Uruk that the Sacred Marriage is instituted and developed. A high priestess will take part in this ceremony, thereby representing the goddess. The Sacred Marriage is a rite linking the fertility of agricultural land and of people with the sexual power of Ishtar, originally known as Inanna, with the former depending upon the latter. Inanna in this ceremony marries either the king or a high priest of the deity Dumuzi. Such sexual encounters are initiated by the woman. Although based in Uruk shrines, this is a public rite, one deemed vital for the good of the community, an occasion for joy and reverence for those taking the place of Inanna and Dumuzi.

In cosmic images of stars and skies, of mountains and paths, of gates and doorways and caves, they make metaphors to carry the weight of their myths. Humbaba is akin to Scorpio, whose celestial gate lies at the crossing of the Milky Way and the ecliptic, and whose doorway in the mountains therefore, through which Gilgamesh passes to help his dead friend Enkidu, is guarded by scorpion men. In defeating that ogre of cedar forests the two men gain access to the celestial world of death, located around the south celestial pole. In defeating me, the Great Bull of Heaven, who is Taurus, the two men gain access to the upper celestial world, located around the north celestial pole. When Enkidu throws my severed haunch at Inanna, he is throwing it at the sky.

I am Taurus, rising with the sun at vernal equinox. To my left is another conspicuous constellation, made of four corner stars, one being bright blue-white and another lurid red. There is a belt cinched across the middle of this stellar man. It is Gilgamesh, who in Babylon will be named URU.AN.NA, the Light of Heaven, and who in Athens will take the name Orion, the Hunter.

As they watch the sun cross the sky, they define the world they live in – the real world and their metaphorical world. On the summer solstice the sun rises at its most northerly position, making its greatest arc across the sky, while on the winter solstice it rises at its most southerly position, making its smallest arc. The point halfway between these two positions they call the equinox, one of which occurs in spring and one in autumn. These four positions – moments in time passed once per year – are akin to the four cardinal points of space: east, south, west and north. They are fixed, dependable markers of the world, giving them a framework in which to live their lives, in particular to follow their agricultural practices. In ritual and rite they acknowledge this framework, celebrating the cycles of the natural world at various times, including at the spring equinox, which they name Akitu, their new year. At this time, before the onset of the dry season, they harvest their crops. Thus I am associated with spring, with harvest and with renewal.

The stars too mark temporal locations through their year, and some stars and stellar clusters are especially important to them, such as the Heliades. Because their knowledge is rooted in the oral tradition, passed down from generation to generation, it has lasted for aeons: stable, certain, unchanging, encoded in myth and story for all to hear and understand. It is a mythic tapestry of tale, couched in terms of the night sky and those who journey across it, of the stars and their mysteries, of Sun and Moon, and the five other wanderers. Yet the planet they live on rotates around an axis, and the direction that axis points in changes over a period of twenty-five thousand years. This axial precession means that the position of the stars on the horizon at times of dawn and dusk changes over time, introducing a discrepancy between knowledge provided by myth and reality.

At dawn, when those observers of the heavens look at the stars, they see that the constellation they expected to see, memorialised in myth and known for aeons, remains buried below the horizon. This is a circumstance of fear for them. Such discrepancies tell them that something terrible is occurring, bringing contradiction, discord, even conflict. Their stellar framework is changing, and it frightens them. In response they make new myths. If the stars are out of place at sunrise, that is a destruction by fire: if they are out of place at sunset, that is a destruction by water.

In times long before the glory of Uruk, the equinoctial marker for the rising of the sun in the constellation of Taurus was seven stars grouped together, the Heliades, situated above a constellation formed as a flowing river of water. As Uruk was built and made glorious by the people of Sumer, that marker moved, to the consternation of many. New myths were spoken, reflecting these changed circumstances, myths describing a calamity in the heavens.

I live at a particular location in the night sky whose celestial coordinates are known. There is a calamity in the sky around me and I lose part of my body, my hindquarters. New myths are spoken of me in which my body is separated into two pieces, fore and hind. In this way the frightening facts of axial precession are acknowledged, comprehended, then memorialised into new culture – new myths, accompanied by new rites.

Four horses pull the sun chariot across the sky, each a cardinal point in time marking a position in the year. If the horses run amok and the chariot falls, maiming or killing the charioteer, that is ancient knowledge changing to fit new circumstances. They are always telling tales, these people, that the metaphorical frameworks in which their minds find refuge reflect and commemorate the world around them. Experience is transmuted into story century by century. As the planet's rotational axis precesses, so do their tales progress.

I am the Great Bull of Heaven, once whole and ferocious, with snorting nostrils and a baleful red eye, brought down to the middle world to fight, but laid low by two men of Uruk. I am dismembered, my haunch thrown in contempt at the sky. Inanna weeps. I remain a bull, but am changed by my experiences, as are those who look up at me from their terrestrial observatories.

I am Taurus.

Chapter 5

Knossos

They approach me as I charge at them, youths all of them, male and female, walking, though not at speed. Their faces bear a variety of expressions, some excited, some exhilarated, some showing signs of fear. At the last moment before impact they reach out with both hands, grabbing my horns and leaping up, using their momentum and my horns to perform a somersault over my back and land behind me. There is a crowd watching who roar in approval and clap their hands, for this is a public display. Waving at friends in the crowd, the youths gasp for breath, jogging this way and that as mild winds off the Aegean Sea ruffle their hair.

I am the bull of Knossos, where rich Minoans live amidst painted splendour. On the island of Crete I am surrounded by the Mediterranean Sea, by mountains and plains, by bronze, by plaster, by amphorae of wine. On the extensive fresco painted upon a wall of the Central Court at Knossos I am light brown in hue, my long horns pointing forward, my neck a solid arc of muscle. My hooves are yellow, my tail whisking high in the air. They wish to tame me through such perilous sports, but I am a wild animal.

It is their custom, this bull-leaping, in which I am forced to participate. The athletes are all young, each of them at the boundary between child and adult, for this is a rite of passage essential to community life. Both male and female wear loincloths, their hair knotted and curled upon their crowns, longer locks falling down behind them. Their muscles are delineated with dust and sweat in the bright Mediterranean sun.

42

The rulers of Knossos use the bull-leaping ceremony to strengthen their claim to sovereignty. Though I am fierce and dangerous, they raise me and my kin in special locations, some of us to be led into the bull-leaping arena, some of us intended for more mundane uses. Thus the rulers direct how Minoans right across Crete view both myself and the Knossos elite. I am the vehicle of their ambitions. Though other palaces contain bull-leaping courts, it is at Knossos where the most experienced priests live, where the best athletes are trained, where I and my kin are raised and where the most spectacular ceremonies are held. I am complicit in the rites in which I participate, to the benefit and continuance of the rulers of Knossos.

But to leap over me is dangerous indeed. I charge at them with all my strength, my head lowered so that I can use my horns to gore them and toss them away. Dust scatters in clouds from my hooves as I charge, and I snort and groan. But peril is important to them. That their young lives are threatened by me is part of the ritual, a metaphor of the role of chance in their lives or of the will of the deities they believe in. They see me not only as a real bull but also as a symbol of the wildness of the natural world. I am their external danger, pushed outdoors throughout their age of bronze.

To leap over me is also a public act. These youths live in tightly-knit communities, some with high status, some with low. But theirs is a communal life whatever their social standing, with events such as bull-leaping part of an intense public ritual. Not only must these youths pass across the boundary between child and adult, they must be *seen* to pass it. I validate that passage.

To leap over me symbolises the end of childhood and the beginning of adulthood. Bull-leaping is a metaphorical door into a perilous new world, where many new feelings, rites, deeds and responsibilities must be managed. The Central Court at Knossos is the theatre for this ceremony, and so to that place I am led, there to wait, to see and listen, and in time to participate.

They vary how they leap. Sometimes they vault over my torso as I run, but more often they attempt to somersault over the entire length of my body from horns to tail. Another variation is to jump up just before I strike them, stretching their legs out in front of them, but this is exceedingly difficult, as it is when they try the same move in the pike position, the upper part of their body bent against the lower.

Sometimes the rite is different. On occasion they will have me lie down with my legs tucked up underneath my body, then leap over me in safer style. Sometimes two aides hold my horns that the ritual be safer still. They also lead me to their stone altars, upon which I am able to raise the front part of my body by clambering up using my forelegs. This is not easy, but I perform anyway.

Of course, they do not always succeed in their athletic manoeuvres. It gives me great pleasure to chase scared youths out of the court, and even to knock them to the ground, but there are always people nearby ready to pull bruised youths aside, or even to try to calm me. Yet when I am enraged I am difficult to calm. They see the wild, fierce, unpredictable animal then. In turn, they do not attempt to dispatch me during these rituals, since the purpose is to make a spectacle symbolising their superiority over me.

They who live in Knossos are obsessed with me. On great vessels called *rhyta* with two openings for pouring they depict me, sometimes making the entire vessel in my image. Usually a *rhyton* is a horn-shaped vessel with my head at the bottom. They carve the best *rhyta* out of serpentine, decorating them with shells, and even with gold and red jasper. Such vessels are not meant to be used in domestic life, they are ritual objects, unable to hold liquid until plugged, for their bases are full of holes. Other *rhyta* are carved with scenes of bull-leaping. The images of me depicted on these vessels are beautiful and naturalistic, unlike the mannered images of their deities and leaders.

A *rhyton* can hold the ichor of a sacrificial bull. From my dying kin, blood is collected, to be cast upon the ground as a libation. In other formal settings they sit to eat my flesh following the sacrifice, a bull's head *rhyton* integral to the ritual, from which my lifeblood is poured. This is the symbolic repetition of my sacrifice, manifested for the community.

These Cretans did not invent their bull symbolism, they assimilated it from many Mediterranean sources, some going back to the pastoral aeons before cities, before writing, before civilisation. I am ancient. I remember when ice locked northern lands in a freezing embrace, and I remember a time when wild boar, mountain mouflon, ibex and aurochs roamed broad plains, unfettered by domestication. I recall the people who lived in Egypt performing bull-leaping ceremonies, also the people of Canaan and Assyria. Those early pastoralists felt insecure upon their lands, vulnerable to drought, to famine, to the vagaries of storm and flood. Thus in myth and athletic performance they sought to demonstrate their control of the natural world, of that wildness lurking at the heart of their most ancient stories. To control, they tame me.

For they are not now all of a kind in Crete's urban communities. In Ice Age tribes theirs is a tight, intense kinship, each individual the equal of others, only wisdom or shamanic skill separating one from another and placing them into a position of higher status. Yet the people of Knossos do not follow such traditions. The elite use bull-leaping to confirm and strengthen the social order which pleases them. In their increasingly stratified urban world, more rituals are required with which to underline the natural, inevitable quality of their dominance. Such ceremonies therefore become performance, essential public maintenance of their control. The very frescoes they paint illustrate the importance of such rituals.

It is not only frescoes, however, which carry images of them leaping over me. High status Minoan people wear gold rings, usually carved with scenes in which I feature, especially bull-leaping. I am the symbol at the heart of their metaphorical world.

They call their method of applying pigments to plaster fresco painting. Such pigments are water-based, made by grinding various natural colours then dissolving them in the purest water available. The plaster surface acquires a matte surface ideal for displaying the artist's skill. A fresco is durable, and can last for centuries. Some millennia later they will experiment with painting upon a wet plaster surface, leading to better quality work.

Two parts plaster are placed into a suitable container, to which one part water is added, the resulting mixture stirred with a spatula until it acquires a smooth texture. Lumps are pressed out with the flat end of the spatula. This mixture is then applied to the wall using a trowel or similar implement. Smoothing can be achieved with a roughened surface fitting into one hand. The plaster dries quickly in the Mediterranean air, and after an hour is ready for painting. Typically roughs of the image to be painted have been created either on or off the wall, so that the final fresco appears exactly as planned.

Before paint is applied, the surface of the plaster is dampened, a technique used to minimise smears and to limit the spread of plaster dust. Their paints are limited in hue, typically earthy colours and soft greens or blues. I am depicted in browns and yellows against a blue background. Bare white plaster is also used in designs. Dolphins they depict in blue and brown against a pale background. They also use black for the long, curling hair of young women. It is never easy to rectify mistakes, but some amendments can be made with a damp swab.

They walk on tiptoe, making as little noise as possible, but they do not know where they are and they do not know where I am. I live at the heart of a stone maze whose intricate details are impossible to remember, whose winding corridors, identical corners and sudden dead-ends fool even the cleverest of them. They approach me, though they do not know that. I can smell their terror on the air, for they are close now, the heat of their bodies and their movements causing eddies to flow along nearby corridors. I guess there are fourteen of them, seven youths and seven maidens, sent to Knossos from Athens and thrown into this maze every nine years. I am starving. But I wait for them to close, that my attack be swift, terrible and total.

I am a man with the head and shoulders of a bull. Minos Taurus they name me, the Bull of Minos: the Minotaur. I am terrible, ravenous, monstrous. My thews are muscle and sinew, as tough as the trunk of a pine tree. My arms bulge with muscle and my chest is broad. My neck is thick, supporting my bull head, upon which magnificent horns are set. My eyes are red and gleaming, and my nostrils snort the very vapours of the pit. I am impossible to escape.

Minos of Crete was one of three sons born from the union of Zeus and Europa, a deed in which Zeus took my form. Europa's husband was Asterion, king of Crete, who ministered to the boys as if they were his own. When Asterion died, however, it was unclear which of the three sons should become king: Rhadamanthus, Sarpedon or Minos. But it was Minos who in due course ascended to the throne, having first dealt with his rival brothers, which he did in a cunning manner, announcing that he had support from the gods themselves, and thus authority enough to be crowned king. He then declared that he would prove what he said by praying, thinking and then naming a thing for which he wished. The gods, he said, would do as he requested.

Soon enough he took it upon himself to perform this deed, sacrificing to Poseidon and praying that a bull would appear from the depths of the sea. In a loud voice he vowed that once the bull appeared he would sacrifice it to Poseidon. Without delay a perfect bull, the Cretan Bull as it is known, appeared out the sea, showing that Minos had the ear of the gods themselves, and even their favour. Thus he became king, banishing his brothers.

King Minos, however, did not keep his word. The Cretan Bull was so magnificent he kept it for himself, sacrificing a lesser beast to Poseidon. Poseidon was infuriated by this opprobrium, and vowed to punish King Minos for his hubris.

So it was that Minos' wife, Pasiphae, began to feel herself attracted to the majestic bull in her husband's herd, and even to conceive a passion for it. The bull from the sea became her unnatural obsession. Plagued by this inflicted desire, she sought the help of Daedalus, the greatest and most original craftsman of his age. After thinking awhile, Daedalus built a hollow wooden cow coated with a real hide, setting it on four wheels so that it could be moved. He then placed Pasiphae inside, moving the wooden cow into a nearby field, in which the bull with which she was infatuated grazed. There she mated with the Cretan Bull, and from that union I was born.

However, when King Minos saw me he realised the truth about Pasiphae's betrayal, and as punishment enslaved Daedalus and Icarus. Yet he left his wife alone. Indeed, Pasiphae cared for me, called me Asterion, the starry one, that being my true name, and fed me while I was a boy-calf.

As I grew up though I became wilder, fierce and furious, and my mother was unable to feed me, or even to care properly for me. Because I was half-man and half-beast I could find no food fit for me, and this led to a terrible chain of events. I began to view the people of Knossos as my food. I felt rage in my heart and gnawing hunger in my belly. I was confused, lost, lacking a place in Knossos: without respect, kindness, or even anything to eat. I felt like an outcast. In rage, confusion and horror I preyed upon the people of my home city.

Meanwhile, in a bid to conceal the truth about his wife's deeds, King Minos commanded Daedalus to build a grand maze in which to house me. This place of intricate corridors was known as the labyrinth, a construction as complicated inside as the palace at Knossos. Yet during its building the king discovered that his only son with Pasiphae, Androgeos, had been killed by Athenians jealous at Androgeos' skill in the Panathenaic Games. Blaming the Athenians for the death of his son and thus for the ending of his family line, he sailed against Athens and demanded that they pay a price for his son's death. In a terrible voice ringing out across the city he insisted that the Athenians pay a tribute to Crete of seven maidens and seven youths every nine years.

Thus it was that I was thrown into the labyrinth, there to live out the remainder of my solitary life. Yet I was not entirely alone. Every nine years I would hear footsteps along the corridors of the maze. Drool escaped my mouth then, and my stomach began to churn. I would eat soon.

So I wait for them to close, preparing myself for an attack swift, terrible and total. I cannot now conceal the noise of my gurgling stomach, nor quieten the sound of my breath snorting out of nostrils and mouth. I stand up, waiting for those seven youths and seven maidens... yet there is something different this time. Nine years ago, and eighteen, the noise of feet was a pitter-patter rush, sandal soles padding against the labyrinth's stone corridor floors. Now it sounds more like the footfall of one person.

Filled with rage, I leap forward, turning a corner, hurtling along a corridor, then turning a second corner. There I see one figure: a man, tall, strong, wearing the armour of an Athenian, his manner wary, alert, yet proud also. In his right hand he holds a sword, while in his left is a ball of red thread, one end of which trails off into the corridor behind him.

"Who are you?" I say as best I can through my drooling, bestial mouth, my deep, resonant voice reverberating around the maze in which we stand.

"I am Theseus," he replies. His voice is light, absurdly childish in comparison with my roar.

"I am Asterion," I say, "the starry one, Minos Taurus, the Bull of Minos, known to all in Knossos as the Minotaur. For nine years I have waited for fresh tribute. I will devour you."

Theseus raises his sword. "I will kill you," he replies.

We fight as two men, but I have the advantage of stabbing horns, two of them to his single sword. Yet although I am strong and sturdy I feel weak from hunger, while he is fresh, well-fed and full of guile. Moreover, he is a skilled and experienced warrior. We leap at one another, slashing with horns and sword, taking wounds, retreating, then clashing again. He steps aside to deliver a sideswipe while I duck, then raise myself and lunge with my horns. He takes a stab wound to the arm, while I am cut by the tip of his sword across my chest. The stink of blood begins to permeate the air, with human sweat, beast sweat, and the rank odour of my saliva. Yet for all our stabbing and slashing neither of us can gain an advantage over the other.

In the end, it is my weakness which is my undoing. He has endurance, this Athenian warrior, with food in his belly and passion in his heart. I sense some other force standing behind him, encouraging him, calling out his name; a person, perhaps. Some lover encourages him, and that person I guess to be the one who gave him the ball of red thread. I see that fine, bloody line now as an omen of his success and my demise, for with it he will be able to find his way out.

Exhausted, I trip, and as I stagger back he stabs out again and again, piercing my belly and cutting out my entrails. I fall to the ground in my death throes, my voice rattling hoarse, blood spurting from slashed arteries, guts and vital fluids darkening the dusty stone floor.

I am done, and my sight begins to fade. This is my end. Minos Taurus is no more.

I am the horned heart of this Mediterranean culture. In Knossos, they who live on the island of Crete revere me, sensing my wildness, which is untamed nature outside their civilised, urban homes. I am fierce, I am strong, unpredictable and filled with rage. I will charge them, gore them, knock them down and trample them. I will kill them in my bestial fury.

I am also Asterion, the starry one, the horned bull constellation set with stellar splendour in the night sky. Across the Near East and further beyond I ride the heavens, visible to all, permanent and majestic.

I am Taurus.

Chapter 6

Harappa

They approach me on the plains of the River Indus, in awe of my virility and strength. They are small in comparison, slender, with only two legs on which to move. I have four: a quartet of pillars. Their skin is dark, their hair black and unadorned. I am pale with horns on my head, towering above them; and much stronger.

They worship my strength, imagining me as an icon. When they pray to Indra, the warrior deity of the skies, they imagine him as a humped bull – a sacred bull called Vrsabha, that is, the male entity, or Uksan, that is, a bull of age between five and nine years who is soon to be fully grown. I am majestic to them, upon the plain and in their imaginations.

The mountain is tall, standing apart from others in the region, and shaped with extraordinary regularity, its four sides arranged like a pyramid and covered with ice and snow. They call it Mount Kailash, and nobody is permitted to climb it. There lives Shiva, the auspicious one, who with Brahma and Vishnu forms the supreme triad of the people of the Indus and beyond. Shiva is master of the cosmic dance, and also the destroyer. He is an ascetic, and also master of beasts. He is worshipped through the *lingam*, the symbol of divine male generative energy. I guard Mount Kailash, and as guardian have the responsibility of keeping the sacred mountain pure and isolated. Only Shiva and Parvati may have their abode there.

I am Shiva's *vahana*, his vehicle, in the form of a bull. In this guise I am named Vrsabha, though in later aeons they will call me Nandi. As Vrsabha I act as a mount, carrying Shiva wherever he desires to go. I am snow white, symbolising purity and justice. I symbolise the inner mind which is focused upon Shiva, gaining wisdom and experience. I am the guru within, the source and inspiration for the process of dedication to Shiva.

They live upon the alluvial plain of the Indus, in its valleys, and in the valleys of the monsoon-fed rivers around the River Ghaggar-Hakra. It is a region of broad extent and many people. Far, far away lie the civilisations of Ancient Egypt and Mesopotamia, where I also live as the sacred bull. The people of Harappa, and of other cities such as Mohenjo-daro, trade with the people of Sumer and elsewhere in Mesopotamia, as well as with the people of Egypt. Animals live in profusion alongside them: elephants and tigers, wild asses and desert foxes, monkeys, snakes and turtles. Through the air many birds fly: the parakeet and peacock, hornbills and cranes, upon bright water shelducks and teals.

In Harappa I am revered. The fortified city is large, home to more than twenty thousand individuals, a planned urban settlement of baked mud brick houses, all of them built with long-standing expertise, for the people of the city descend from earlier agriculturalists. The management of water is a particular skill, and they construct supply and drainage systems, that Harappa stand as a safer, better resourced city. They are a people who worship and feel reverence for the world in which they live and for the universe, whose metaphors are Brahma, Vishnu and Shiva. In special buildings they worship the deities of the divine triad. There are many other deities, however, a profusion of them, each representing a particular area of life. Like all preliterate cultures or cultures for whom the advent of writing has not brought significant religious changes, this polytheism indicates that the archaic way of using nature as metaphor and framework has not been superseded. Polytheism emerges because people living in oral cultures use nature as their medium, as their inspiration and as their memory. Because nature is experienced in many different forms, each aspect can be linked to a particular part of life. Nature is used as a template for those who cannot write and who therefore must use memory aids in order to keep essential knowledge alive in their communities.

Anthropomorphism is a style of thinking deriving in part from their inability to see beneath their perceptions into the plane of reality. Some of this follows from their lack of understanding of the less obvious processes of nature. Yet such processes – the noise of thunder, the power of lightning, the ash and roar of a volcano – *must* be explained, this being a fundamental part of their mental functioning, which makes a mental model of the world, a model that must be *coherent*. The need for such explanations, however, is not the entire reason for anthropomorphism. Aspects of nature are anthropomorphised in order to provide a foundation for parts of their lives. If a story can be told of a volcano deity, it makes the memory of volcanoes and their character – their history in the lives of the people too – easier to articulate. All this can in turn be woven into stories which everyone can remember. Their understanding of those parts of nature upon which their lives depend – animals, plants, weather and land – is in fact exceptional. Such understanding, however, is couched in terms of themselves, in order that, through speech alone, and over countless millennia, vital knowledge never be forgotten.

Though they live in a world of myth and ethereal realms, they are perfectly capable of using reason-based thought. They have exceptional knowledge of natural history and of the use of materials to make implements and tools. When hunting, they adhere to irrational superstitions, yet the knowledge they use in order to track prey is entirely reasonable. They are subtle, complex thinkers with deep minds.

They are skilled with their hands and eyes, these people, devising ritual objects of clay. They have an appreciation both of the necessities and the arts of life, using clay with perception and flair; they also know copper and bronze. On rotating wheels they make pottery, decorating it with abstract and animal symbols. They wear garments of cotton dyed with various natural colours. They use writing and make many images. They keep cattle, including me, and also many fowl. Their houses and larger buildings are made from mud bricks, their construction skill being a wonder for me to behold.

Though their buildings are complex, the ingredients for a mud brick are in fact simple: soil with up to one half clay, sand, straw or conifer needles, and water. To make a mud brick, the soil and water are first mixed. To this, sand is added, before the straw is mixed in. The mixture is poured and squeezed into a wooden brick mould then left to dry for about five days. To protect the bricks from direct sunlight, which might create cracks, they cover them. Once removed from the moulds, each brick is allowed to dry for another few days before use.

In Harappa they live complex lives, each of them confined to a social stratum. Some of the people in lower strata are not cared for as well as those enjoying higher social status, and these communities often know mistreatment and disease, including leprosy. From me they also know that consuming disease of the lungs which infects many Harappans. They know battle also, suffering life-threatening wounds inflicted by weapons. Their civic organisations are strictly managed, leading to a conspicuous lack of idiosyncrasy and deviation across the Indus valley culture. Each city, however, has its own particular culture and traditions, for there is no central authority controlling the whole civilisation.

Trade out of Harappa is undertaken with produce and textiles, but also with semi-precious stones such as lapis lazuli and carnelian. From the south of the subcontinent they obtain gold and copper, which they use with all their natural skill and dexterity. In Mesopotamia they keep small colonies which function as trading centres, that trading be constant and secure.

On intricately carved seals made of soapstone they show Shiva seated, and they also show me, viewed from the side, my horns long and held vertically with pride and majesty, my chest broad, my haunches tough and muscular, my tail hanging down to the rear. Other seals show beasts of the region, such as the elephant. In the earliest incarnation of the city of Harappa some seals show the trident of Shiva, as well as the demiurge depicted in seated posture, one leg folded beneath him, one extended.

The tales they tell describe me as the son of Shilada the ascetic. In order to receive a blessing from Shiva, Shilada endured a grievous ordeal, that he might have a child with immortality and the blessings of Shiva. Thus it was that he fathered me. I grew up to become a dedicated follower of Shiva, submitting to several of my own ordeals in order to become both his *vahana* and protector. Yet it was from Shiva's wife Parvati that I received my greatest wisdom, that of the Tantra, which in later years I taught to my eight most devoted followers. Those followers I sent out to the eight most distant parts of the world, that they disseminate my knowledge to others; in due course they became the originators of the doctrine known as Nandinatha Sampradaya.

Another tale they relate describes my struggle with Ravana, the demon king of Lanka and the great adversary of the Ramayana. Because he behaved like a restless, agitated forest monkey while waiting for Shiva, I cursed Ravana, declaring that his kingdom would be burnt by a monkey living in the forest. Lanka was duly burned by Hanuman, just as I predicted.

A third tale speaks of an altercation which occurred between Shiva and Parvati. During a lecture given by Shiva on the subject of the meaning of the Vedas, Parvati lost her concentration, much to the distress of Shiva. In response to this lapse he incarnated her as a lowly fisherwoman, that she make amends in suitable style for her distracted mode of listening. I realised that it was up to me to reunite Shiva with the wife he truly loved, so I took the form of a great whale, a beast of the ocean who without delay began to harry and batter the people amongst whom Parvati, as fisherwoman, lived. After a while, exhausted by the scourge of me-as-a-whale, Parvati's father declared that whichever man had the strength and courage to kill the whale would marry his daughter. So Shiva took the form of a local fisherman and killed me in that incarnation, that he return Parvati to her true form, as his beloved wife.

I sit in every shrine of Shiva. I am the great bull upon a platform, lying on my belly with my legs folded beneath my body, my horns removed, my hump prominent, decorations wrapped across my back and down my flanks. In this seated form I look towards the inner sanctum of the shrine, that most sacred space of Shiva, protecting it. Representing the ethical goal of every individual, I remind worshippers that their minds must be directed towards Shiva and his teachings. I am pure white, my four legs representing love, truth, righteousness and peace.

In many temples south of the Indus they portray me with the head of a bull and four hands, holding an antelope and a battle axe; sometimes carrying a mace and *abhayamudra* also. Sculptures of me in human form may be found beside the doorway of many a temple to Shiva. Such sculptures show me akin to him, a crescent moon set amidst tangled hair, the third eye in my forehead. Often, two of my hands are pressed together in adoration of Shiva.

In holy Varanasi I am free to wander the streets, revered and protected by those who live there, branded on one flank with the trident sign of Shiva. They believe me to belong to Shiva, and so in those streets I am respected.

I am frightening, like Shiva. Strong, relentless, wise and filled with zealous energy, I follow my master, dedicated to him. It is natural to be frightened of me, as it is of Shiva.

I am Vrsabha, and I have ancient heritage.

I am Taurus.

Chapter 7

Babylon

They see me in the northern night sky, the heavenly bull whom they call Guanna: in their form of writing, GU.AN.NA. In Babylon I am already ages old, alongside UR.GU.LA, the lion, and GIR.TAB, that which claws and cuts, the scorpion. In Mesopotamia they knew me as GU.GAL.AN.NA, which is to say the Great Bull of Heaven. Aeons in the past, before Babylon, before even Uruk, they saw us three in the night sky and formulated our constellations, visualising our shapes in the patterns of the stars. In those ancient times they used our trio of constellations to mark three of the four cardinal points: the winter solstice, the vernal equinox and the summer solstice. We three were fundamental to their agricultural calendar, but I was listed first.

I am the bull of heaven who marks the point of the vernal equinox. In the Babylonian star catalogue known as the MUL.APIN the stellar cluster Hyades upon my cheek is called the Crown of Anu, who is Lord of the Sky. The Pleiades above my shoulder they see as a clump of bristles. I trace my roots into half-remembered times before written history. I am ancient, with long memory.

They relate how I am the first husband of Ereshkigal, the goddess of the underworld, who is elder sister to Inanna. When Inanna is rejected by Gilgamesh, she implores Anu to allow her to lead me down to the terrestrial world, there to bring justice. But I am killed by Gilgamesh and Enkidu.

Yet I remain a deity to they who live in Babylon. I am the servant of Anu, sworn to do his bidding, for he is the ruler of all the deities, master of the heavens, of creation and of justice. I am Anu's envoy. My wife Ereshkigal is the Queen of the Earth. She rules the land of the dead, devising and enforcing underworld law. I am her servant too, playing my part in administering natural justice so that order and peace lie across the lands. I am the instrument of deities, doing their bidding without complaint or deviation.

Ereshkigal and I have a son named Ninazu. He epitomises peace, sympathy and healing, being of mild and empathetic character. In these lands the people know that the spring equinox is the time of annual bounty and revitalisation; to heal a body is to restore its capacity, and that my son does with most excellent skill. He acquired those attributes from me, of course, for I regenerate, bring health, bring fertility and the power of divine justice also. I am owed reverence, for I am huge, strong, ferocious and courageous, my two great horns a symbol of my natural power. All rulers look to me in their efforts to acquire status and the awe of their subjects. Kings wish my majesty to be transferred to them.

Babylon is a feast for the eyes. Temple towers vie with tall trees for dominance of the skyline, with shrines and royal palaces clad in blue-glazed tiles. There is bronze, there is silver and there is gold, but only for the richest nobles. The city walls are built to last, tall and massive, atop them a road.

The date palm is common across the region, but also found in gardens and growing wild are willows, tamarisk trees and poplars. Bulrushes rustle on many banks of the Tigris and Euphrates. The black-and-red glossy ibis is a common sight, along with sand grouse, herons and the white pelican, with more ordinary birds such as rock doves also in residence. Gazelles and goats can be seen everywhere, less so the marauding jackals.

In the time of Nebuchadnezzar there is one building for which Babylon becomes famous, the ziggurat of Etemenanki. Nebuchadnezzar has the following words inscribed upon it:

As to Etemenanki, the ziggurat of Babylon, of which Nabopolassar, king of Babylon, my father, my begetter, had fixed the foundation – and had raised it thirty cubits but had not erected its top, I set my hand to build it. Great cedars which were on Mount Lebanon in its forest, with my clean hands, I cut down, and placed them for its roof.

The city sits on a broad plain at the eastern bank of the River Euphrates, its walls forming a square inside which the ancient settlement lies. A moat protects it from invaders. In the eastern quarters many temples are built, all of them from older times, alongside quays for merchants' boats to dock and unload their wares. Babylon is a hub for trading, famed for its wealth and majesty. The central quarters are laid out on a grid pattern, the so-called Processional Way passing along its centre. This route is important because it leads to the Ishtar Gate, west of which lie two vast palaces. Embellishments can be seen everywhere: dragons carved in relief, and images of myself too.

The temple of Marduk is glorious to behold, a square construction with specially manufactured gates of bronze. A central tower rises high into the air, composed of eight progressively smaller sections atop which lies the temple. These sections can be climbed by the outer spiral steps winding their way upwards. Halfway up, seating allows worshippers to rest before attempting the remainder of their climb.

With wedge-shaped marks in soft clay they make lists of constellations, of bright and colourful stars, and of the five wanderers, the planets. They are highly skilled at cuneiform, mostly using clay to make impressions upon, but also wax and even ivory. The best cuneiform is deemed to have artistic value, bringing joy through the beauty of its arrangement alone. More often, however, it is workaday, its value in recording agricultural labour, ownership of fields, produce harvested and bread baked.

The great temples also make much use of cuneiform, declaring in writing who they are and what they own.

Cuneiform is written with a reed stylus, the rear end of which has a circular cross section, the fore end being a wedge of various shapes. Holding the soft clay tablet in the left hand and the stylus in the right, they make rows of impressed marks, each one a narrow triangle, using circular impressions to denote numbers. This writing position means certain angles of wedge are difficult to make, and so their texts use a range of shapes easy to make with the stylus held in the right hand. Their sexagesimal numerical system uses an accumulation of symbols for one, two, three, four, five, ten, twenty, thirty, forty, fifty, sixty, six hundred, three thousand six hundred, and thirty-six thousand. Each tablet tends to flatness on the reverse, the obverse side tending to convex. Tablets can be larger than a handspan, but most are smaller. They begin writing on the flat side, and when that is full turn to the convex side, in this way preserving the marks first made. Beginning at the top left corner they make their marks downwards, starting the second column when the bottom edge of the first is reached. When the flat side is full, they turn the tablet over and begin at the top right, working downwards as before, with later columns moving to the left.

Tablets can be altered if they are not baked by adding water to moisten the clay, but permanent records are baked to preserve the records they hold. Tablets baked with fire turn dark grey or even black.

They have many scribes in Babylon, the overwhelming majority being boys or men. Schools exist for the training of novices in the writing of cuneiform, a process in which a teacher inscribes on one side of a tablet a fragment of a poem, of the Epic of Gilgamesh, or perhaps some local proverb or technical term, which the novice scribe then attempts to copy on the other side. Their efforts are usually worse – I know writing is a difficult art. But once trained, a good scribe can work in the royal court or be the aide of a city governor; such would be a high-status position. Lower status destinations include places of shipbuilding, of textiles manufacture, or of transport. The majority, however, work in agriculture, recording harvests, the quality and state of irrigation channels, and the number and functionality of agricultural implements.

In Uruk, I recall, their signs were more like pictures – pictograms as they are termed. As Uruk faded and Babylon grew and became great, those pictograms became more abstract, turning into shapes made from combinations of wedge marks. In some, a ghost of the original shape remains. In later centuries, as writing became even more complex and widespread, they realised that the up-down columnar method of writing was laborious, and that if they turned their marks and the direction in which they made them around by ninety degrees it was easier. So a cultural shift occurred. Although they still made partitions, as with the old columns, their writing was now efficient, abstract and accurate. Yet as I saw in Uruk, the change from oral culture to literate pushed their societies in a particular direction, to their great loss, and to their deprivation as a whole. Their world will not be the same again. Moreover, they will never look at me in the way their distant ancestors did.

So they observe the Sun and watch the Moon, understanding with subtle knowledge how they and the planets travel across the night sky. Two thousand years after the beginning of writing those who live in Babylon collate all that they know, creating clay tablets known as Three Stars Each. In these works, various stars are named. Some say their origin lies in Sumer, some say in Elam.

Just as their cosmos is divided into three tiers, the heavens, the terrestrial world and the underworld, so the celestial sphere is divided into three. Enlil rules the northern third, with the equatorial third belonging to Anu and the southern third given over to Enki. In the work Three Stars Each, all twelve months of the year have three stars, giving thirty-six in total. These star trios are named MUL, which in earlier times was a pictogram of three stars. When referring to the Pleiades, the cluster of six stars lying just above my shoulder, they use the name MUL.MUL. This is an additive process of nomenclature.

Two centuries later they revise and update their knowledge using improved observations, formalising the work they call MUL.APIN, in which I am listed at the head. This work includes the three archaic constellations GU.AN.NA, which is myself at the vernal equinox, UR.GU.LA, the lion at the summer solstice, GIR.TAB, the scorpion at the autumnal equinox, and also SUHUR.MAS, which is the horned goat at the winter solstice. We four mark the cardinal points of the annual cycle of the heavens. We are ancient and we are venerable.

The other eight are called MASH.TA.BA, that is, the great twins, AL.LUL, the crayfish with pincers, AB.SIN, Sin's daughter who wields the seed furrow, GISH.ERIN, whose scales mark the balance of fate and fortune, PA.BIL.SAG the defender, GU.LA who is the great lord of the waters, SIM.MAH who are the fishes with tails like swallows, and LU.HUNG.GA, who works in the fields. The course of the Moon is also listed in the MUL.APIN.

In later centuries of Babylon they observe that the sun no longer rises amidst my stars upon the vernal equinox. For over two millennia I have been their anchor at that time, but now the location is taken by LU.HUNG.GA. I am half a bull, my haunches removed; yet I remain great.

They write, these people, and it makes them understand one another in a form different to that of speech. Yet just as their languages mutate over time, their letters, though permanent, do change in the making as many years pass by. The turning of the great cycle of the year is a circular path, the end of the year attached to the beginning of the next, but the evolution of letters takes a different route, and as decades become centuries they alter their mode of writing – a one-way path.

In the beginning, I was at the head of it all: the great bull. I was shown as a triangle or a square with two projections raised up – that is, as my head when viewed from the front or the side, and if the latter with a dot for an eye. In some places I was named Aleph: the ox. This pictogram represented a sound made during speech: a phoneme. In Ugarit, five hundred miles from Babylon, they later decided to use a cuneiform version of the older pictograms, creating an alphabet of thirty symbols, each representing a phoneme. This alphabet was being used one and a half thousand years after the development of writing in Mesopotamia.

It is the Phoenicians, living on the Mediterranean coast a few hundred miles west of Babylon, who develop a stable alphabet which spreads across the world, and in that alphabet I remain at the head. My pictographic form is turned around by ninety degrees so that my horns point to the right. Aleph is still my name, my phonetic value corresponding to the first vowel, that is, *a*. A few centuries later the Phoenicians give their alphabet to the Greeks, who turn my head around a further ninety degrees, so that my horns point downwards, the top of my head sloping up from the lower left horn. Later, that slope is corrected so that I become symmetrical, and am called Alpha. Those Greeks at the beginning of their use of the Phoenician letters write one line from left to right, then the next from right to left, and so on downwards in a style known as boustrophedon, which is to say, how the ox ploughs.

From bull's head pictogram through semi-abstract symbol to fully abstract A, my head has stood at the beginning of the alphabet for the entire duration of its life. I am the leader. When they write A, they are depicting my head, horns and all, upside-down.

My strength, ferocity and potency are depicted in the stars for all to see. On the ecliptic I stand, proud upon the path of the zodiac, my horns delineated in stars, my eye shining red. Guanna I am, one of their twelve constellations, who will endure through the ages and be passed on from culture to culture. I am forever, I am the bull of heaven, whose face and horns stands at the front of their system of writing – the leader. I am majestic upon the land, huge and heavy, and I remain so even when pulling their ploughs. My strength is known across the Near East and on into other, more distant lands. Guanna they name me, but I am something far older.

I am Taurus.

Chapter 8

Memphis

They have many priests in Memphis, and one, a man of experience and wisdom allied with divinely inspired insight, identifies me amongst the kine around the city, nurturing me throughout my life. They take me into a temple, where I am housed with honour, my stall comfortable and filled with everything I need. During my lifetime I am revered as Apis, the living incarnation of Ptah, ancient deity of Memphis, creator of the universe, deity of artisans and folk of craft, husband to Sekhmet. They worship me in their temple, the flawless bull, and at the end of my life I am taken away for the mummification ritual. The priests deem me perfect; the correct bull for the rite. They anoint my body with unguents and wrap me using pieces of cloth. Then they take me to my tomb.

I am Apis, sacred bull, known across the Old Kingdom. Son of the deity Hathor, I am the power and vitality of life, a force of creation, but because I am so strong I am also a symbol of war. I am spoken of as *ka*, a word which also means life-force, and which they believe to be one of the three ineffable parts of an individual: soul, *ka* and *ba*. Therefore I am potential. I am life creativity, I am virile.

They depict me on a carved stele, dedicating their hieroglyphics to me: to Apis. I am shown from the side perspective, seated, my legs tucked beneath my body, my torso stained red, my head left as the straw colour of the stone. My horns, however, are shown from the front perspective, a wide, majestic crescent, in which a solar disc is placed.

They make many metaphors, these people of the Old Kingdom. They believe the *ka* of an individual carries on after the death of the body, and is then able to reside inside a stone statue. This form of thinking is ancient, however, even at the origins of the Old Kingdom, and I can trace similar customs back in time, through the spirits in the mud brick walls of Çatalhöyük all the way to those in the stone walls of the caves of Lascaux.

I am the active incarnation of Ptah, able to think for myself and perform deeds. No passive servant undertaking orders, I may act on my own behalf in order to change or direct the affairs of the Old Kingdom. In later centuries under different dynasties I am the living incarnation of Osiris.

In vaults beneath the Serapeum of the stepped pyramid of Saqqara they bury me and my kind inside large stone sarcophagi, we who dwell inside temples and are worshipped by all. At the Serapeum there are at least sixty such tombs, built over many, many centuries. Before the Pharaoh known as Ramesses II each tomb is constructed as a separate space with a shrine above it, but afterwards they lay us to rest in underground warrens connected by tunnels.

They also revere my cow mother, Hathor the sky goddess, patron of joy and love, whom they sometimes depict as like them except with a cow's head or ears. The sun disc is my mother's symbol, and few deities show it. I, her son, am one such.

Founded by King Menes, who unified Lower Egypt and Upper Egypt, Memphis was originally called White Walls, most likely from the use of white-painted bricks on Menes' palace, although pale limestone could be another possibility. That stone covered the exterior of the three pyramids built centuries later at Giza nearby, dazzling all who saw them in their heyday.

I see that Memphis is vast, with many necropolis buildings lining the Nile's western shore. A great canal links the city to the river, for commerce is vital, Memphis thriving on all sorts of trade. Palaces and temples stand in abundance, while gardens also are common. Palm trees sway in the breeze, while along the river nearby innumerable papyrus plants cover the riverbanks, making a thick green border. There are acacia trees here too, elsewhere cypress and the sycamore fig. Between some of the papyrus clumps the fabled blue lily grows, the flower sacred to Egyptian customs because of its scent and mild psychoactive effect.

Birds are common throughout this region. Black and white cranes perch atop high walls, while in winter months crescent-winged swifts fly in all directions, their distinctive cries echoing across the city. Falcons prey on smaller birds. Around Memphis many animals live: camels and donkeys, cobras and mongooses, hyenas and crocodiles.

Memphis is the focus of the worship of Ptah, creator deity but also linked to artisans, smiths and other makers. There is one enormous temple dedicated to Ptah, but also smaller shrines dedicated to his wife Sekhmet and his son Nefertem, along with other deities. Bastet is the cat-headed goddess – the local sand cats are domesticated shortly before Memphis emerges. The importance of this city to the Egyptians is such that many pharaohs use the cliffs nearby to excavate their fabulously decorated tombs.

They write with hieroglyphics, sacred carvings, using many signs to set down their meaning. It is a mixed system, some symbols representing ideas or whole words, also known as logograms, some representing a sound or sounds, known as phonograms. Some hieroglyphs may be recognised as images of objects, often an animal such as the ibis or serpent, but the interpretation of such images is not necessarily direct. An image of a hand for instance does not relate to hands, it is a phonogram representing the sound *t*. Hieroglyphics are a context dependent writing system, in which signs can have more than one function; but such is the way with many writing systems. An additional idiosyncrasy is that the vowel sounds, a, e, i, o, u, are almost never represented in hieroglyphics.

An alternative spelling of my name is Hapi-ankh. Some hieroglyphs are based on the body parts of animals, including those of my kind. The people of the Old Kingdom have a symbol, the *ankh*, which is like a cross in which the upper section has been replaced by a loop, and which they use to mean life or being alive. This symbol is often used alongside two others, the *was*, which is a sceptre and which they use to mean dominion or power, and the *died*, which is a pillar, that they use to denote stability. Some say these three symbols are based upon certain of my body parts, which the people of the Old Kingdom believe transmit semen, the vital male ichor responsible for life creation but also linked with control and power in society. Bones they believe to be source and conduit; bones which are white, like semen. Therefore some say the *ankh* is one of my thoracic vertebra, the origin of semen, the *died* is my sacrum linked to lumbar vertebrae, that is, fused bones between the hip bones of my pelvis, and the *was* is my penis, through which semen emerges into the world. In this way they conceptualise both what is sacred and important in their societies and the symbols used to represent such concepts.

In Ancient Egypt, the papyrus plant grows in profusion along the banks of the River Nile, especially in the delta region. It is a water-loving species, which the locals harvest in order to make papyrus scrolls, upon which they write. Papyrus has tall, firm stems with a triangular cross section, inside which lie multiple fibrous strands. These strands are cut into narrow strips, many of them side by side, with the same on top but at an orientation of ninety degrees. In this way, many layers opposing one another are built up into a papyrus sheet, which can then be moistened and pressed. Left to dry in the baking sun, the sticky sap within becomes a binder, gluing the sheets together. To make a scroll, sheets are battered into their final thin composition, then pasted together.

Papyrus is ideal for use in hot countries since it is resistant to deterioration, stable and relatively inexpensive to manufacture. Its lightness compared with stone tablets or even vellum makes it a favourite during transportation, although when rolled it can crack, and if the atmosphere is moist it will acquire a veneer of mould. As the centuries roll by, however, it becomes a writing and drawing favourite across the Mediterranean region.

I was not always so high in reverence amongst the people of the Old Kingdom. At first, as the civilisations of the River Nile began to urbanise, I was perceived as the son of Hathor, important in rites dedicated to her, rites in which I was sacrificed and reborn. It was only later that I became an emissary and agent of Ptah, then of Osiris, then of Atum.

In the earliest days I was deemed sacred as an animal, not as a deity. They observed my strength, my ferocity and my ability to fight, and from such qualities made me worthy of worship. Because kings required these attributes I became a metaphor for kingship, a symbol of the righteousness of the monarch's rule. Kings often assumed the title 'Strong Bull Of His Mother Hathor'. There were two other bull cults in Egypt, but in due course all were merged under the worship of Hathor. I then was revered as Apis.

Thus in and around Memphis I become the agent of the city's deity Ptah. I am therefore an incarnation of the king, his living symbol, imbued with the same attributes. The kine of the fields around Memphis are in the main black in colour, with white markings on various parts of their bodies. When upon the death of the old Apis a priest seeks a new sacred bull, he looks for a particular set of markings: a white moon crescent upon my right flank, a white diamond upon my forehead, a white mark on my back in the shape of a vulture wing, and a mark underneath my tongue shaped as a scarab beetle. I must also display twinned tail hairs. When all these marks are seen together on one calf they know it to be sacred, the special one – myself, Apis.

Swiftly I am taken to the priest's temple, riding in a beautifully decorated golden boat built for me. Inside the temple I am housed in luxurious accommodation along with a group of cows. I am worshipped: I am sacred. But that is not all. The cow who is my mother is believed to have conceived me from a heavenly ray of light, so she also is treated with reverence and named the Isis cow; and when she dies, her burial follows sacred traditions.

They use me as a diviner, able to tell the future, or at least to offer advice on it. My movements around the accommodation in which they hold me are taken to have meaning, which they pore over, hoping to gain wisdom. Even my proximity to them brings benefits; they believe my strength can transfer into them. They also believe my breath has curative properties. Yet I am not entirely a bull of the temple. My accommodation includes a window to the outside, through which ordinary folk can view me. On holy days they lead me through the streets of Memphis, flowers and jewels decorating my flanks, my shoulders and my head.

But as with them, as indeed with all animals, a day comes when my body fails and I die. My death marks the beginning of a series of special rites, encoded into their beliefs, written upon many sheets of papyrus. In the early days of the kingdom they mummified me standing up on my feet, that I be set upon a platform of wooden slats. In due course the New Kingdom is founded, and that is the time at which they bury me in the tombs of Saqqara. Soon, the subterranean complex of chambers that they call the Serapeum is constructed in order to house me and those who follow after me. Me and my kind are fixed in writing now, the dates of our acquisition, reign and death all recorded, sometimes with the names of our mothers and the places where we were born. We live on in crowded lines of reverential letters.

I have an afterlife in other cultures too. When the ages of the Pharaohs are over, new rulers from Persia take over, followed by men from Greece. Their influence is strongest in Lower Egypt. The Greeks wish to merge the principles and deities of the Egyptians with their own, that a sacred harmony be created in which Greeks and Egyptians stand and worship as one. Thinking that Amun might be suitable, Alexander the Great tries to forge that unity, but he fails. His general Ptolemy I Soter tries a different tack, realising that animal-headed deities are unsuitable for Greek worshippers. Since Amun is more popular away from the Lower Kingdom, and observing the respect offered to Apis right across Egypt, Ptolemy devises a new image for me, which he names Aser-hapi, that is, Osiris-Apis. Some time later they begin calling me Serapis.

Of course, I am not shown as a bull, nor as a bull-headed man. I am fully anthropomorphised, a man, akin in appearance to the Greek deities of the underworld. I have a basket upon my head with which to measure grain, that which they name *modius*, a symbol of the underworld. At my feet, however, lies a symbol of rulership in Egypt, the *uraeus*, which is in the form of a serpent. I carry a sceptre, also indicating kingly authority, while the three-headed hound of the underworld Cerberus lies at my feet. With my wife at my side echoing Isis I am acceptable to Egyptians and to Greeks, and in this form I live on as Serapis for many centuries.

They turn me into a sacred bull because they see their kings through my eyes, kings who must epitomise the greatest of great attributes. I am virile, I am huge and formed of solid muscle, I am ferocious and I am strong. A king should be all these things. Through the ages I have been a deity to them, a member of the pantheon of all parts of Egypt, from the flooded Nile Delta to the hills and desert sands of Upper Egypt. Thus cow and bull are intertwined, as they who worship me are intertwined. But there is more. I am a metaphor of their culture, a symbol – themselves, yet in different form. From ancient sources I am come, walking on hoofs, great horns upon my head, between which I hold the solar disc of my mother Hathor.

I am Apis, sacred, beloved, born, dying and reborn.

I am Taurus.

Chapter 9

Canaan

They worship me as an idol. A golden calf I am called by some, but I am a golden bull, and great beast descended from aurochs I shall always be. From gold they make me, that I appear before them as a graven image, worthy of reverence, shining untarnished and pure; yellow gold in the light of the yellow sun. They are their own people with their own culture, but in a bed of eastern Mediterranean cultures they will always sit, imbued with the sacred bull, patrolled by the sacred bull, overseen by the sacred bull who is me, omnipresent, vigilant and powerful. I am the mount of ancient Canaanite deities under whose aegis these people live.

As a statue of gold I am pure and noble, alloyed with the otherworldly essence of past ages. I symbolise strength and fertility. Deities and I... we both demand worship.

In buried ruins at the port city of Ashkelon they find a small image of me, three and a half thousand years after it was cast, an object made of bronze to which other metals have been applied. Given that well over three millennia have passed, this image of me is in extraordinary condition, its legs, tail and ears all present, and one horn too. Yet the temple in which it was set was destroyed when Ashkelon submitted to violent conquest by the Egyptians. These observant archaeologists date the time of manufacture by referring to associated pottery finds, deducing from strata and nearby material when the image was made. They are amazed to discover such an unexpected find. Indeed, they are delighted.

This image of me is neither simple nor naive. Just over four inches in length and about the same tall, it weighs twelve ounces or so. The maker manufactured its various parts separately: legs, ears, tail and horns all fit into prepared sockets of the correct size. The legs and the head are made of silver, while the horns and the tail are fashioned from copper. Made of bronze, the torso's surface is well polished when complete, burnished so that the alloy more resembles gold. Inside, however, they place lead in order to increase my mass, that in holding me weight equates to religious significance.

Yet I and the deities who ride me are not sacred to all. The Israelites make great efforts to ridicule and bring down such Canaanite reverence, denouncing he who some name Baal. Even Ephraim, the great prince, esteemed in Israel, is put to the sword for worshipping Baal. Kissing images of calves is deemed a sin by the Israelites, who wish to declare themselves an independent, separate people in the region. There is strife and there is battle.

Canaan has a few small cities, but most of its people live in villages. This is a rolling, undulating country, with settlements made in and around the hills. Higher, more mountainous regions are often terraced, making them unsuitable for settlements. Most people are farmers of one sort or another, with herds of sheep, goats and cattle most popular.

In Canaan they name their ruling deity El, whose full name is Thoru El – that is, Bull El. In the language of those living in the city of Ugarit *thor* is the word used to describe me, which equates to *shor* in the tongue of the Hebrews. El therefore is symbolised by me and my image. Small figurines of El made of bronze are common, the deity portrayed sitting down so that the figurine can be attached to a wooden pole and carried in processions.

In the early days of the Israelites they also manufacture and worship such images of El. At Dhahrat et-Tawileh in the hills of Samaria, which lies between Galilee and Judaea, there lies a mountain ridge at which a holy site is built, circular in plan and a little elevated, around which a low wall stands. It is a cult site set in free open air. There, the people worship me in bronze form, that tiny alloy figurine of a sacred bull. Made with high technique, it is cast according to Canaanite metallurgical traditions. It is El, visible as me, the sacred bull: *El, who brings them out of Egypt, is like the horns of a wild bull for them.* (Numbers 23:22)

There is strife now between the Israelites and those more northerly peoples from whom they wish to separate themselves. A sacred bull is something divine and holy, yet in later texts the Israelites refer to the golden calf. Is that the real me? Or is such writing the propaganda of the Israelites as they make themselves a secure and independent people? Such texts are collected in the Bible, the sacred book of countless people yet to be born. The golden calf is indeed a deliberately mocking image, shrinking me to young form in order to show that such idols are immoral. It is denigration by metaphor and by word. In the Book of Kings it is written that as Israel separates itself from Rehoboam, Jeroboam, the first king, makes a pair of holy sites in which he places two golden calves, his intent to pit himself against the sites of Rehoboam's capital city of Jerusalem:

> So the king took counsel, and made two calves of gold. He said to the people, "You have gone up to Jerusalem long enough. Here are your gods, O Israel, who brought you up out of the land of Egypt." He set one in Bethel, and the other he put in Dan.
>
> (Kings 12:28-12:29)

But this is a form of propaganda, Jeroboam's deed described in the Bible text shown to readers as if it is wrong. Yet at the time, the Israelite priests would not have viewed the deed so negatively; they would never have used the word calf. They would have described these idols of me as bulls. Diminishing a graven image in such a way is intentional mockery.

Other speakers made mockery in a similar way. Hosea says:

Your calf is rejected, O Samaria. My anger burns against them. How long will they be incapable of innocence? For it is from Israel, an artisan made it; it is not God. The calf of Samaria shall be broken to pieces.

(Hosea 8:5-8:6)

Those men writing Biblical texts could never reconcile themselves with idols supposed to represent a deity. Bible polemic is aimed at such practices, not least the worship of El the bull in northern temples. The God of Israel is deemed to be something higher, greater, more abstract.

I, then, am not a golden calf but a bull. In those nascent, perilous times for the Israelite people, why did they require a graven image of me after Moses went up to the mountain to receive divine wisdom?

It is written that many days passed, and still Moses did not reappear. The frightened Israelites, huddled together with enemies on all sides, lacking water and anything to eat, needed reassurance from something they might trust – something ancient, something with roots. Aaron the brother of Moses was the temporary leader of this nation without a homeland, but he lacked his brother's moral weight, and he struggled to keep order. The people spoke their demand: *Make for us an elohim to go before us.* With Moses missing, they wanted something to take his place – an image presenting them with sacred surety, that they not lose their connection with the divine. Facing this unstable situation, Aaron felt he had no choice but to accede to their demand. There was even the possibility that some Israelites would return to Egypt: a disaster.

Aaron and his people have spent many long years in Egypt, so they are steeped in Egyptian images. In Memphis and elsewhere they have seen me as Apis, whom the Egyptians name Hapi: the sacred bull. This is a truly ancient image of worth and high ancestry. A real, solid bull is therefore required, and Aaron knows this.

He recalls what he knows of the traditions of Egyptian priests relating to Hapi:

When upon the death of the old Apis a priest seeks a new sacred bull, he looks for a particular set of markings: a white moon crescent upon the right flank, a white diamond upon the forehead, a white mark on the back in the shape of a vulture wing, and a mark underneath the tongue shaped as a scarab beetle. It must also display twinned tail hairs. When all these marks are seen together on one calf they know it to be sacred, the special one – Apis. Swiftly it is taken to the priest's temple, riding in a beautifully decorated golden boat built for it. Inside the temple it is housed in luxurious accommodation along with a group of cows…

In his mind Aaron sees me as the Apis Calf, symbol to all Egyptians of hope, of renewal, of longstanding continuity. He senses the importance of finding a new beast after the death of the old one. In his mind therefore a synchrony is made: the nervous Egyptians are like nervous Israelites. He realises he must reassure them with something they know already, a symbol of hope, a calf which in due course will grow to be a mighty, noble bull. That it is made of gold echoes the golden boat in which I arrive for the Egyptians.

Moses is gone for forty days before he returns to find the Israelites worshiping the golden graven image. That they celebrate this idol as Egyptians celebrate the new Apis Calf must have been a moment of horror and shame for him. He smashes the two tablets upon which his Ten Commandments are written. Yet there is a chance left for proper atonement.

I am the sacred bull of this region, beloved of many, revered, spoken of in ancient poems, carried in the minds of many people, manifested in word and in deed. In shrines I am set as bronze figurines, holy in metals of high regard, manufactured with care and devotion. I have silver and copper legs, ears, tail and horns, being high-born and worthy. Time is required for my manufacture since my creation is no small deed. I receive veneration and homage, being cherished and admired.

I am Taurus.

Chapter 10

Rome

They perform the *Taurobolium* ceremony in public, sacrificing me to the deity Magna Mater, Great Mother, who in older days was called Cybele. Almost two hundred years have passed since events in Jerusalem initiated the cult of Jesus Christ, and Rome is at the centre of the pre-eminent empire of the world. In Italy, in conquered Gaul, in Hispania and desert-bordered Africa they request the good wishes and concord of Magna Mater through my sacrifice at the *Taurobolium*. In this way, they ensure the welfare and good fortune of both the state and its people.

I am great and worthy. To sacrifice me is to propitiate those in heaven who have an interest in Rome and its empire. I am a symbol of what must be done to appease the gods, or, at the very least, to attract their attention.

At the earliest *Taurobolium* they use curved weapons to sacrifice me. The design varies little, being a straight blade with a sharp point made of metal, a second, hooked blade poking out just behind the end. Sometimes they call this a *harpe*, sometimes *ensis hamatus*. I am struck first with the straight blade, an action that stuns me and which opens up my flesh so that it bleeds. Then they rip the wound open with the second blade, completing the deed. Yet this Roman rite is different to other, more ancient rites, which over long years have settled in my memory. From the time of the Ice Age I have been approached with respect, allowed a moment in which to agree with whatever is going to happen – that I be hunted and killed for food and sinews, or that I be sacrificed. These Romans do not allow me that moment. They are domineering and zealous. They know what they want, and they get it.

Yet the ritual is popular amongst Romans, who view it as an important part of their identity. They are keen to expand their empire, these people, to rule distant kingdoms, to exploit resources there, to take captives and make them slaves. The *Taurobolium* is one part of this colonial voraciousness, which they use to tell themselves and all who look upon them who they are. They dispense justice according to their ancient rites. That bundle of birch sticks known as the *fasces* symbolises authority and justice. Usually containing an axe, and bound by a red leather ribbon, it is also used in the physical dispensation of justice, since the sticks can be used to whip and flog, the axe to execute.

In essence the *Taurobolium* is a rite of purification supporting and improving the moral lives of the Romans. They believe in deities above and around them, led by Jupiter and Jove, who provide the environment and the heart of their morality. In the early *Taurobolia* it is a single individual who is the recipient, the ritual of my sacrifice being dedicated to them. That individual receives beneficial blessings, of purity, of good health and of long preservation. For twenty years such blessings last, whereupon they may be renewed through a new *Taurobolium*. They call that day the *natalicium*, which is to say, birthday, showing how they see the ritual as a kind of rebirth; the individual is described as being in *aeternum renatus*, that is, eternally revivified. My blood regenerates them, making them better able to resist the damage and misfortunes of the world. My blood brings lasting health and good preservation against disease and affliction. My blood lasts for two decades.

Perhaps then they remember the distant past when my blood was spilled to bring rain after drought, when my blood revivified flowers, and even grain. Their harsh, domineering empire after all is built on many earlier cultures, all of whom sought life and survival through me. It may have been a wolf that suckled their city founders, but it was my kin who accompanied them through their lives. For important individuals the *Taurobolium* is performed, but it can just as well be for the Emperor, for his family, for the military men who are Rome's muscle, for the Senate, or even for the city itself. A *Taurobolium* can be for all Romans.

These Romans have inherited a world of iron. Bronze is a good metal which can be worked in pottery kilns reaching a temperature sufficiently hot to melt tin. However, tin is not always in good supply, and there are other reasons for desiring something better. Bronze is not a strong enough metal for certain purposes. Thus, although known about and used in tiny quantities for well over two thousand years, steel from iron ore becomes the metal of choice. Steel is harder to make, however, requiring a special furnace to reach a high enough temperature in which to smelt iron ore; but it is harder and more durable than bronze. The metallurgist's craft too is more complex. Not only do impurities in the haematite ore have to be removed, adding carbon to make steel from iron is a tricky process. But once iron begins to take over from bronze, the new metal and its smelting practices spread swiftly. The Romans delight in steel.

For well over two and a half centuries the *Taurobolium* is performed, but the ritual does not remain constant; rather, it evolves alongside changes in the character of the Roman Empire.

At first the ritual is akin to more ancient rites rooted in the lands of Mesopotamia, of the Nile and of the Levant, it being my sacrifice – blood shed in ritual moments. I am washed, removing mud and accumulated dung from my body, along with insects and parasites. Cleansed, my short hair is revealed in all its glory. Then they decorate me with woollen ribbons coloured white or red, and a disc is affixed between my horns. Over my broad back they place that fringed blanket known as a *dorsuale*.

I am now ready for sacrifice. Only the one known as the *Archigallus* is allowed to perform the rite. He takes a handful of grains and scatters them over my head, following this by dripping water on my forehead, between my eyes and horns, in this way signifying that I am an animal of sufficient quality and merit. Then he pulls a few hairs from my head, throwing them to the fire nearby. Responding to a signal, the man known as the *victimarius* takes his weapon and with a single blow stuns me, tearing my flesh at the neck with the straight blade, then opening it fully with the curved blade so that my sacrificial blood flows free. Thus, I die. In a hand bowl this hot, redolent blood is caught and taken to the altar, that it be presented to the deities. To show that the sacrifice is true and holy they splash my blood upon the altar in the manner known to all who perform such ancient rites.

For more than a century this sacrifice continues, but then it begins to change. Though the essential actions of the ritual are the same, they use new words to describe them, indicating a new focus. The bowl in which my blood is caught is now named the *cernus*, a word originating in the Greek *kernos*, which is to say a clay pot with small receptacles around its rim. This ritual object in times before the Romans was used to burn incense and sometimes offerings, the receptacles holding many symbolic things, from resin or wine to seeds or water. Its use by the Romans shows that my blood is inherently sacred – a gift, no less, from the Magna Mater herself. They use their word *vires* to imply holy powers carried by my lifeblood. So it is that my blood is received and passed to the altar: *accipere* and *tradere*.

For almost a century this ritual is performed, but then it changes again, into a form quite unlike the original. Now not only is it I who am washed and cleansed, it is the recipient of the blessings too. That individual puts on a fresh white *toga*, then is to walk towards a pit dug into the earth above which wooden planks are arranged. They call this pit the *fossa sanguinis*. Once beside the pit, the individual descends into it, the planks rearranged above them so that they lie deep and concealed. I am pulled towards the pit, to stand above it so that my head and foreparts lie above a hole in the planks, whereupon the *Archigallus* scatters grain and water, then pulls hair from my head and burns it. At a signal, the *victimarius* stuns me with his straight blade then rips open my neck with the hooked one, so that my blood spurts out, pouring through the hole, that the individual beneath be soaked in it. A little of that flow is caught by the *Archigallus* in the *cernus* bowl, then taken to the altar as before. When I am dead and the rite is almost over my body is dragged off, clearing the way for the planks to be drawn aside so that the recipient of the lifeblood blessings can clamber out and walk free. Thus it is that the *vires*, those powers of blood spurting from my neck, are transferred from me to the recipient, giving them what I once had: blessings of longevity, of revivification, of renewal.

At this time there is competition between the deities of the Romans and the single deity of Christianity. Immersion in blood during the *Taurobolium* mirrors immersion in water during baptism, a development perhaps intended to improve the stance of the old deities in the eyes of a public tending towards the new religion. My blood is a life-giving fluid, just like water.

They believe that in their world Mithras walks. Mithraism is a male-only cult, and highly secretive. Unlike the public *Taurobolium*, its rituals, including bull sacrifice, are private, albeit known through carven images and writing. Seven levels of initiation into the cult are each associated with a specific meal undertaken in strict ritual circumstances. Mithraic temples are common in Rome itself and across conquered Europe, but far more rare in Egypt and regions of the Levant.

Most likely Mithras came from Persia. He is a deity of light, truth and honour, born out of a rock, the central figure of the cult of Mithraism and popular across much of the Roman world. In addition he is associated with merchants and the protection of warriors. He appears in the lives of the Romans a century or so after Jesus, disappearing from view and from written histories three hundred years later. Yet there is no great body of tales about him as there is with Mars or Venus, so that all which is known is sourced either in texts dealing with the cult of Mithraism or depicted in carven images.

A few details are well known. Following his sacrifice of me for instance, Mithras wanders the world, meeting the Sun, whereupon the pair shake hands and eat various portions of my carcass. He is not always depicted alone when adventuring. Sometimes he is accompanied by a pair of torchbearers, Cautes and Cautopates, sometimes by Oceanus, who is the manifestation of the sea, and occasionally by a lion-headed individual.

I, however, am the animal at the centre of Mithraism, once again to be sacrificed. They call an image of Mithras the Bull-slayer the *Tauroctony*. In this fashion they disseminate and manifest the prime mysteries of the cult of Mithraism, which is so cryptic and elusive, yet so widespread. In dank, dark caverns they practise these mysteries, the chambers so narrow I could never myself enter, were that ever required. The underground hollows of the cult are indeed gloomy and cramped, but, since there is no central rite as with the *Taurobolium*, that does not matter. What matters is the image: Mithras and me, my blood revitalising grain, greenery and flowers.

The earliest image of me and Mithras is very old, made little more than a century after the appearance of the man known as Jesus. This *Mithraeum* is a dedication to the slave bailiff called Alcimus, he who laboured in the service of Claudius Livianus, the praetorian prefect employed by Emperor Trajan.

Every *Mithraeum* of the Roman world features a scene of *Tauroctony*, which is central to the symbolism of the cult. We are entwined, Mithras and I. He kneels upon me, holding me down so that I do not ruin the rite with my struggles, grasping me with his left hand at my nostrils as he looks out, in his right hand the dagger which stabs me. I am bleeding, providing that lifeblood which is at the heart of the ceremony. A serpent and a dog lick this blood, while a scorpion grabs my testicles with its claws. I writhe and lurch and bellow, but Mithras holds me down so that I cannot escape. Other symbols surround us. My tail is composed of a sheaf of corn. Though he and I struggle inside a cavern, above us lie woodlands, and above them the Sun and the Moon. From the Sun to the left a ray of light penetrates down, illuminating Mithras and the raven which sits close by. The Moon is placed to the right. At each corner of the image stands one of the four winds, blowing from the four cardinal points, while various scenes of Mithras' life lie to either side. Some carven images show him ascending into the skies in a chariot.

This cavern in which we struggle is a depiction of the underground chamber in which the members of the cult of Mithras perform their secret rites. Often the cavern has one central aisle, to either side a raised dais. Usually fresh water flows nearby, either in the form of a stream or sometimes from a spring, with basins used to transport ritual waters. The cavern itself does not stand alone, since most often there is a small antechamber at the entrance, with additional rooms inside for the preparation and storage of food.

They slay me in ritual places because they believe my blood is sacred, revitalising them one by one, bringing fertility to their lands, preserving the lives of favoured individuals, of groups, or even of the entire Roman Empire. I am at the heart of their rites, giving my lifeblood in ceremonies of ancient origin which come from all corners of this part of the world. I am their symbol for revivification, for preservation and good health, for the fertility of green things and also of themselves. They *need* me.

I am Taurus.

Chapter 11

Pamplona

They keep me inside a pen so that I am ready to charge down a street. Constrained, I snort and stamp my hooves – the very essence of ferocity and wildness. They have a rite in mind for me, one in which we shall both participate, they on their thin legs and small feet, me solid in muscle and massive bones, my hooves clattering. We will run down the street together. It will be a wild ride, flailing arms and legs, screams and shouts, horns lowered to gore, breath snorted out of wide nostrils.

The running of the bulls pits them against me, side by side; and I shall give no quarter.

It is the end of September, and large numbers of local worthies, farmers and merchants are gathering in the town of Pamplona. A religious ceremony commemorates the martyr San Fermin at this time, Catholic in nature, enacted eleven hundred years after the crucifixion of the one they name their saviour. As centuries pass by this ceremony merges with the local fair, but considerations of weather mean the date is moved to a more clement week in which the holy day of San Fermin is placed: July 7th, the *Dia de San Fermin* as it has been since the end of the sixteenth century. After this change, amendments are made to the ceremony, including the Procession of the Giants and Bigheads, until it becomes world famous as the Pamplona Running of the Bulls, one of the largest *fiestas* of the year.

They call it *encierro*, which in their tongue means confinement. In the early days of the ritual the residents of Pamplona would block street entrances and exits, thereby confining me to a fixed transference route, a practice evolving from the need to transport my kind from fields to the city. As often happens in such cases, young men attached themselves to the practice in order to display their courage, a tradition which in due course became the modern ceremony. Those narrow streets down to the *Plaza de Toros* – the bullring – became known in due course as the bull run.

It takes about two and a half minutes for me, my kin and the locals to run the full length of the route of nine hundred yards. The beginning lies in my pen at Santo Domingo: the end lies in a bullring. Of course, the ceremony does not always go according to plan – this is a ritual of wildness and rampaging chaos. Sometimes the *encierro* descends into disorder, and lasts thirty minutes or more.

They begin by singing in Castilian and in the Basque language – Pamplona is in Navarra, in the far north of the country – their text a benediction to San Fermin, the city's patron saint, to whom they also offer prayers for guidance and protection. This recitation occurs three times and is sung by all the participants. The streets are filled with their white shirts and trousers enlivened with red belts and scarves. None may drink liquor, however nervous they might be. Spectators watch through open windows or from the balconies of tall houses, having risen for the occasion; a while before this gathering, the sun also rises.

At eight o'clock in the morning a rocket is set alight, its explosive crack the signal that the gate to my pen in the Old Town has been opened. Yet it is not quite time for me to charge. Soon, however, I and my kin are loose, whereupon a second rocket explodes high in the air. There is noise now in all directions, from the sky, from the buildings around me, echoing up and down the narrow streets. This noise enrages and frightens me, my instincts urging me to run.

Four narrow streets await me and the fifty or so other bulls. We charge at top speed, the participants before and between us, a chaotic, headlong rush down to the *Plaza de Toros*. I care nothing for them, these small creatures on white legs, whose arms flail and whose heads bob. They run like chickens, hither and thither, while we black bulls charge amongst them. Many of them fall in the rush, and some we trample, but I care nothing about their injuries. Sometimes one will come close, and by chance I have a moment to lower my head and gore. The injuries I give are deep and bloody. Sometimes I kill them. But I also am in danger. These Pamplona streets are cobbled, and I skid along them helter-skelter, at once infuriated, agitated and daunted.

Street cobbles become slippery with blood. We black beasts stumble sometimes, as do the runners. All is mayhem, all is a frenzy. Yet while they have a certain bravado, I have rage. I, after all, am a scion of the sacred bull.

In the end I face a crowd of people in the bullring, amongst them *matadors*. I am stabbed with blades, I am beaten all over my body, I am goaded until fatigue makes me weary and I begin to stumble and fade. With an *estoque* the *matador* ends my life. I hope for a clean kill, his *estoque* piercing my heart. If it pierces my lungs I will drown in my own blood.

In Andalucía they face me inside the bullring, a wide open space surrounded by seats in which the crowd sits awaiting the spectacle to come. They are called *matadors*, and they are ready to fight with me. But I have been bred from bulls going back centuries, and I have been trained so that I will attack anything which moves. This is me in full flight with full force: stamping, snorting, charging, ready to gore with my horns, to throw the *matador*, to knock him down, to chase him to the edge of the bullring. I have memory going back thousands of years and I have seen every sacred bull custom, but still I take part in their ceremonies, and still, whether they fight me or stab me or just stand and taunt me, I know they are worshipping me, testing themselves against me, the great bull who will forever be stronger than them. They may have their cunning and metal spikes, but I have the bulk, the muscles and the bones.

I am ancient yet still alive, worshipped in modern times within the bullring, my line continued through uncounted generations. They may deem themselves my superior, but that is because they know I tower above them. My strength alone makes me superior. I will never be broken. I will remain great for all time.

It is the Romans who bring me and my sacred culture to the Iberian peninsula. Local Visigoth tribes take to it with alacrity, continuing the tradition, though it lacks rules and structure. Later, when Rome fades and a new culture arrives in Andalucía, the Almohad caliphs who preside organise and enjoy bullfighting, grasping that the spectacle of me in the bullring is something to be admired. They bring a ritualistic style to the proceedings. Moreover, when they confront me they are mounted on horses. Still later, Alfonso VII de León y Castilla, when marrying his bride Berengaria de Barcelona in 1128, organises many bullfights as part of the celebrations. Such traditions spread throughout Spain, multiplying and changing. In later centuries I am released from captivity into the town square, where the high status man of the locale, mounted on his horse, faces me carrying a lance. As medieval decades pass by, individuals emerge whose role it is to complete the ritual process of my death – their modern version of all those ancient sacrifices. They are named *matatoros*, literally killer of bulls, and to the north of the country they acquire high status and accrue much wealth as a consequence. Some rulers and other nobles take to hiring them for annual festivities.

Then comes a change in the laws. Alfonso X de Castilla publishes and promulgates his *Siete Partidas* statutory code, forbidding various forms of public festivity, including the work of the *matatoros*. He deems bullfighting fit only for the noble classes, making it ignominious to receive payment for such a role. Yet even at this stage, the men who use their worsted capes to aid the man on horseback begin to receive more attention from the audience, who roar and cheer in approval or dismay. Thus does the tradition of fighting me continue through the centuries, slowly evolving, until, much later, more significant changes appear.

It is 1726. In the region known as Ronda a man appears who amends the bullfighting tradition. He is the great Francisco Romero, *matador* of skill and renown, who brings to the bullring a new sword known as an *estoque*. But more than that, he introduces something iconic which will forever be associated with bullfighting: the red cape known as the *muleta*. Because he fights me on foot, not on horseback, these novel modifications begin to spread through all strata of society. People observe that horses and high status are not necessarily required to enjoy bullfighting. The medieval centuries are gone, and Spain is a different country, with more people and a wider range of traditions. Now when they fight me in the bullring they come from high caste and low, always on foot, their red cape at their side. It is the birth of the fully modern tradition.

New amendments continue to appear in their techniques of fighting me. Soon there are supporting men in the bullring known as *cuadrilla*, and bullfighting in this mode spreads to all corners of the nation. Yet as it does a subtle shift in the social whirl around bullfighting occurs, the clamour for public bullfights becoming too loud and strong to be restrained. So the nobility lose their noble art, forfeiting their high place to all living in Spain. This is the beginning of a popularity universal across the country, and with that prevailing social approval comes renown and esteem for *matadors* and all associated with them, shattering the old traditions begun by Alfonso X de Castilla.

As the 1800s progress, further amendments to the bullfighting culture are introduced. Now an event lasting an entire day in which many bulls are fought across a morning and an afternoon, the structure is amended into one half, the *media corrida*, in which six or eight bulls are fought. A quarter of the way through this century the *media corrida* becomes the standard structure of bullfighting across the nation. Ten years later the renowned bullfighter Francisco Montes, sometimes called Paquiro, publishes a document called *Tauromaquia*, in which not only the precise rules for fighting me are set out, but even such details as the size of the bullring. Two decades later the first such bullring is constructed in Valencia, and in the decade that follows the new standard format is set. I and five of my kin will confront three *matadors*, each with a supporting *cuadrilla* team. Thus bullfighting in the modern era begins.

Not all the people of Spain are passionate about bullfighting. Many, however, argue in favour of the tradition, the debate swaying this way and that. Advocates say it is an ancient tradition which should be continued and respected as part of the Spanish nation's heritage; Spain's founding is closely linked to bullfighting. Also, there are economic and cultural benefits, not least because bullfighting entertains hundreds of thousands of people. Bullfighting, however, is never mere entertainment, it is an art form, one expressing the relationship of myself with human beings. I am a sacrificial animal, as I have been for aeons.

Then there are the conditions in which they keep me and my kin to consider. I am no ordinary bull – I am a Spanish Fighting Bull, one of a special breed who have lived for centuries. I live in wilderness specially created for me, wilderness which acts for the good of the world. Moreover, the conditions in which I am kept – to live for good, long years – should be contrasted with the fate of farm animals, few of whom live beyond a couple of years, then to be sent for slaughter in unpleasant conditions. It is true that my meat is desirable, consumed with delight after the bullfight, but there will always be more of my kind ready to fight at the *corrida*: the unique Spanish Fighting Bull.

They begin the fight in the later part of the afternoon, when the great heat of midday is passing or gone. My life and those of five others are to be taken by three *matadors*. A music known as the *paso doble* – literally the rhythm of the march – announces the arrival of the *matadors* and their assistants the *picadors* and *banderilleros*. The *matadors* take all the crowd's attention, their costume marking them out: dark, flesh-hugging trousers, the bicorne hat known as the *montera*, and a jacket of great sophistication and expense. These *trajes de luces* – literally suit made of lights – are decorated in profusion with golden adornments, so that they glitter in sunlight.

They release me into the bullring via the *toril*, which is to say, the bullpen gate. I arrive with flair and showy exaggeration, running, looking around with alertness, then moving once again. I take in all that I can observe, hear and smell. Especially I am looking for anything which moves. Then I see the *matadors* making the traditional *verónicas* move, which is showy, using the large cape – a number of passes intended to mark the introduction of the bullfight. This is what sets me in motion again, stirring my instincts. I must charge!

I hear applause coming from the crowd surrounding the bullring. Such appreciation is proportional to how close my horns come to the body of the *matador*. Other skills deemed worthy of celebration are the calmness of the *matador* in the face of my stamping charge, and how well, and with what fluidity he wields his cape, for it is to that swirling target that all my rage is directed.

I charge by. He twists and turns, swinging his cape in front of me. Enraged, all I see is that moving target. This is my only instinct – to strike down, to gore, to stamp.

Yet it is not only instinct which propels me. They breed me and my kind in special establishments, selecting for the characteristics they desire. I recall this of course from my days in Memphis: in Spain it is not so different. A natural bull will not always charge when required, whereas I and my kin always will; thus are we trained. Such powers lie within us, however; they are not drawn out by ill-treatment of any kind. I am special, as always I was in the histories of diverse peoples down the ages, revered in a multitude of ways by so many different cultures.

Most of my kind are sent to slaughter for their meat. I and my fighting kin live for years longer than they, in due course to be transported to the *corrida*. I will have seen a minimum of four years if I am to oppose a *matador*.

Now comes the next part of the bullfight. In this, the *picadors*, mounted on horses, use their lances to pierce my hide, wounding me, infuriating me, yet weakening me to a considerable degree. I begin to feel wary now, sensing that those swirling capes are not the real danger. They never use more than three lances, these lesser men in their metal leg armour and chamois leather trousers. Their costume is based on silver, not gold. But then I hear a trumpet call. What next? It seems that the Principal of the *corrida* has declared enough. Yet now I am approached by men on foot carrying barbed sticks: the *banderilleros*. They advance upon me, using their skill and dexterity to stab me in the shoulders with their *banderillas* so that I am forced to lower my head. I am in pain, my body failing. I am confused, furious, wary, unpredictable. Yet I retain my horns, most ancient of sacred symbols, which for nineteen thousand years have attracted the gaze of all who see patterns in the night sky and relate them to the world around them. I am still the sacred bull. I can gore, and I will.

Yet now I am wounded and fading. It is the final act of the *corrida*, the *faena*, marked by another trumpet call. The *matador* covers his *estoque* sword with his cape, beginning a series of fluid movements and passes which to all who watch seem like dance. *Matadors* often come from bullfighting families, and such moves – the very art of the bullfight, the display of skill, the manifestation of thousands of years of sacred design – are learned when young.

What goes on in his mind as he strikes and taunts me? I can gore him more easily now that we are in close proximity, and given the chance I will. My kin make most *matador* woundings during the *faena*. Yet he is skilled and practised at his art. His ploy is to maximise the peril to himself while not risking death. After all, my horns can and do pierce *matador* flesh, and not so deep beneath that pale pink skin lie vital organs. If I can kill him, I will. It is my instinct. This is the position in which I have been placed by the people of Spain, but I accept the challenge, the spectacle, the ritual. I remain the sacred bull of long ages.

There are a number of ritual passes made by the *matador* as we struggle. The *trincherazo* takes place with the *matador* placing one knee upon the ground. I also face the *pase de la firma*, in which the *matador* stands still while fluttering the *muleta* in front of my face, and a natural pass, which maximises the peril to the *matador* since in that move he separates the *estoque* and the *muleta*, making himself more obvious to me.

Yet the *matador* is not only focused on me and what I am doing. This is a public ritual, a ceremony watched by a large crowd, as happened ages ago in Knossos where they leaped over my horns. A skilled *matador* plays out his design in front of the crowd and with them always in mind. He is a performer on their behalf, following the ancient rite of working with the sacred bull. Thus, at the end of the *faena*, he makes more play of the danger of my horns, yet even now he is controlling the spectacle, that he remain safe whilst allowing the crowd maximum excitement and satisfaction.

Now though the end is at hand. I am to be killed once again. The bullfighter takes his sword and prepares himself for dispatch – the moment of truth as the Spanish call it. When he is in the correct position and observes that my front feet are pressed together, he aims the *estoque* over my horns and into my withers, that the blade pass straight through me into my heart.

I am killed. I am gone. The bullfight ritual has used me and sacrificed me, the sacred bull, one of that long line of the scions of Taurus. It is a moment of bloody spectacle and ancient ceremony. Through my eyes, now closed, the crowd see themselves: in awe of me, revering me, yet able to control me. In the bullfight I am a metaphor of themselves.

I am them, yet in my own body and with my own strengths. They see me as a stellar bull, awesome, depicted with clarity in a pattern of stars, with great horns pointing forwards and a red eye. Above my shoulder a cluster of six stars lies, while upon my face there is a V-shaped stellar cluster.

In their minds there is identity between the bull of the night sky and the huge aurochs which roam the plains where they live. I am the aurochs displayed in the heavens, marking out a crucial time of year. They use me in order to remember and know their annual calendar. I am the great timepiece of the heavens, rising and setting without fail, recalled in their mythic lore.

They make many images of me. My horns they see as the crescent moon, giving me a lunar association handed down over the ages. They worship me, they capture me, they keep and then kill me. I am their cosmic sacrifice.

Over aeons I move through the heavens with my own motion, so that in due course they split off my rear quarters. Yet I remain worthy of reverence, sacrificed every year, huge, fierce, noble, filled with vitality, strength and power. The new kings of the urban world identify themselves with me and all that I represent.

In their ever-changing legends they remember me, the sacred bull sacrificed in countless rituals for the good of the people, of the monarch, of the state. My blood is eternal vitality, renewing, reviving, preserving. Upon their earth my blood is shed. Innumerable public ceremonies are watched by the people in shrines, temples and other sacred places.

I am the horned bull in the night sky, I am the bull inhabiting building walls, I am the horned standard of bronze upon a wooden pole, I am the Great Bull of Heaven, I am the terrifying beast in the labyrinth, I am the vehicle of cosmic principle, I am the sundered bull of heaven, I am the bull prepared in great houses for sacrifice, I am the Golden Calf, I am the foundation of purity and vigour, I am the fierce animal of the arena to be spiked and speared.

I am Taurus.

Stephen Palmer can be found online at *stephenpalmer.co.uk*,
and at his Facebook page, *Stephen Palmer – Author*.

IFF
BOOKS

ACADEMIC AND SPECIALIST

Iff Books publishes non-fiction. It aims to work with authors and titles that augment our understanding of the human condition, society and civilisation, and the world or universe in which we live. If you have enjoyed this book, why not tell other readers by posting a review on your preferred book site. Recent bestsellers from Iff Books are:

Why Materialism Is Baloney
How true skeptics know there is no death and fathom answers to life, the universe, and everything
Bernardo Kastrup
A hard-nosed, logical, and skeptic non-materialist metaphysics, according to which the body is in mind, not mind in the body.
Paperback: 978-1-78279-362-5 ebook: 978-1-78279-361-8

The Fall
Steve Taylor
The Fall discusses human achievement versus the issues of war, patriarchy and social inequality.
Paperback: 978-1-78535-804-3 ebook: 978-1-78535-805-0

Brief Peeks Beyond
Critical essays on metaphysics, neuroscience, free will, skepticism and culture
Bernardo Kastrup
An incisive, original, compelling alternative to current mainstream cultural views and assumptions.
Paperback: 978-1-78535-018-4 ebook: 978-1-78535-019-1

Framespotting
Changing how you look at things changes how you see them
Laurence & Alison Matthews
A punchy, upbeat guide to framespotting. Spot deceptions
and hidden assumptions; swap growth for growing up. See
and be free.
Paperback: 978-1-78279-689-3 ebook: 978-1-78279-822-4

Is There an Afterlife?
David Fontana
Is there an Afterlife? If so what is it like? How do Western
ideas of the afterlife compare with Eastern? David Fontana
presents the historical and contemporary evidence for
survival of physical death.
Paperback: 978-1-90381-690-5

Nothing Matters
a book about nothing
Ronald Green
Thinking about Nothing opens the world to everything by
illuminating new angles to old problems and stimulating new
ways of thinking.
Paperback: 978-1-84694-707-0 ebook: 978-1-78099-016-3

Panpsychism
The Philosophy of the Sensuous Cosmos
Peter Ells
Are free will and mind chimeras? This book, anti-materialistic
but respecting science, answers: No! Mind is foundational to
all existence.
Paperback: 978-1-84694-505-2 ebook: 978-1-78099-018-7

Punk Science
Inside the Mind of God
Manjir Samanta-Laughton
Many have experienced unexplainable phenomena; God,
psychic abilities, extraordinary healing and angelic encounters.
Can cutting-edge science actually explain phenomena
previously thought of as 'paranormal'?
Paperback: 978-1-90504-793-2

The Vagabond Spirit of Poetry
Edward Clarke
Spend time with the wisest poets of the modern age and of the
past, and let Edward Clarke remind you of the importance of
poetry in our industrialized world.
Paperback: 978-1-78279-370-0 ebook: 978-1-78279-369-4

Readers of ebooks can buy or view any of these bestsellers by
clicking on the live link in the title. Most titles are published
in paperback and as an ebook. Paperbacks are available in
traditional bookshops. Both print and ebook formats are
available online. Find more titles and sign up to our readers'
newsletter at http://www.johnhuntpublishing.com/non-fiction
Follow us on Facebook at
https://www.facebook.com/JHPNonFiction
and Twitter at https://twitter.com/JHPNonFiction